BLM Field Handbook for Mineral Examiners

by Department of Interior

with an introduction by Kerby Jackson

This work contains material that was originally published in 1962.

This publication is within the Public Domain.

This edition is reprinted for educational purposes
and in accordance with all applicable Federal Laws.

Introduction

It has been years since the US Department of Interior released their important publication "Field Handbook for Mineral Examiners". First released in 1962, this work has been unavailable to the mining community since those days, with the exception of expensive original collector's copies and poorly produced digital editions.

It has often been said that "*gold is where you find it*", but even beginning prospectors understand that their chances for finding something of value in the earth or in the streams of the Golden West are dramatically increased by going back to those places where gold and other minerals were once mined by our forerunners. Despite this, much of the contemporary information on local mining history that is currently available is mostly a result of mere local folklore and persistent rumors of major strikes, the details and facts of which, have long been distorted. Long gone are the old timers and with them, the days of first hand knowledge of the mines of the area and how they operated. Also long gone are most of their notes, their assay reports, their mine maps and personal scrapbooks, along with most of the surveys and reports that were performed for them by private and government geologists. Even published books such as this one are often retired to the local landfill or backyard burn pile by the descendents of those old timers and disappear at an alarming rate. Despite the fact that we live in the so-called "Information Age" where information is supposedly only the push of a button on a keyboard away, true insight into mining properties remains illusive and hard to come by, even to those of us who seek out this sort of information as if our lives depend upon it. Without this type of information readily available to the average independent miner, there is little hope that our metal mining industry will ever recover.

Though this volume may not at first seem to be of great importance to gold miners, I feel that those miners with an interest in developing their claims and also patenting them, will find the processes outlined to be of great value.

This important volume and others like it, are being presented in their entirety again, in the hope that the average prospector will no longer stumble through the overgrown hills and the tailing strewn creeks without being well informed enough to have a chance to succeed at his ventures.

Please note that at times it is necessary to rearrange illustration plates in these texts. Any illustrations not found in their original sequence may be found following the index.

Kerby Jackson
Josephine County, Oregon
February 2015

www.goldminingbooks.com

UNITED STATES
DEPARTMENT OF THE INTERIOR
BUREAU OF LAND MANAGEMENT

Memorandum

To: Mineral Examiners

From: Director

Subject: <u>Field Handbook for Mineral Examiners</u>

This handbook is issued to maintain consistency and professional
quality in Bureau of Land Management mineral examinations and reports.
The Code of Ethics for Government Service and the Professional Ethics
of Mineral Examiners should be the guide lines in accomplishing
assigned work.

Many charts and tables in other professional publications are not
included in this handbook. However, significant tables, background,
and legal cases which bring out many phases of valuation engineering
are included for field reference.

This handbook does not restrict the application of all available or
newly initiated methods of conducting an examination. The professional
training and background of the examiner would be used extensively in
making on-the-site opinions and decisions. Good judgment used with
this material will place the results of the examination on a high
plane of accuracy.

CONTENTS

FIGURES

TABLES

CHAPTER I

GENERAL

The Bureau of Land Management

The Bureau of Land Management, guardian of the Nation's public lands and natural resources, was formed in 1946 by consolidating the General Land Office (established in 1812) and the Grazing Service (created in 1934).

BLM is responsible for the conservation, management, and development of some 477 million acres of the public domain, about 299 million acres of which are in Alaska. In addition, BLM administers mining and mineral leasing on other federally owned lands, on former Federal lands where the minerals have been reserved in public ownership for all the people, and on the submerged lands of the Outer Continental Shelf.

The five technical staff activities of the BLM are land management, range management, forestry, surveying, and mineral conservation and development.

Following its objective to encourage the development of the mineral resources on the lands it administers, the Bureau applies the principles of multiple use and sound conservation in protecting the public's interests.

The minerals staff duties include the following:

1. The issuance of patents to mining claims upon application and full compliance with the mining laws and regulations.

2. The determination of the validity of mining claims conflicting with other nonmineral entries under the public land laws, or when requested by other Federal agencies desiring the clear title to the lands for public purposes.

3. The responsibility of providing technical advice and guidance in the appraisal and sale of mineral materials at both the district and State office level.

4. The determination and classification of the mineral character of lands, when the land is being considered for title transfer or exchange under applicable public land laws.

5. The appraisal of patented, or valid unpatented mining claims or mineral lands in condemnation proceedings, either on Bureau motion or at the request of other Government agencies.

6. Providing technical advice on mineral matters required by the Bureau or when requested by other agencies.

7. Cooperation with other Government agencies, State and local governments, private citizens, industry, and the various organizations and associations in the development and uses of mineral resources.

1

The Bureau's objective is to encourage the development of the mineral resources on lands under its jurisdiction in accordance with the laws and regulations. In the exercise of its discretionary authority it will apply the principles of multiple use and sound conservation practices in the following manner:

a. Protect the public's interest in the mineral resource of the public lands.

b. Facilitate the Bureau of Land Management's title transfer and management programs initiated by its range, forestry, and lands activities and support these activities by necessary classification actions.

c. Reduce and maintain on a current basis the pending workload.

d. Cooperate with all users of the public domain, including but not limited to miners, livestockmen, and conservation groups, in support of programs to minimize all types of abuse of public lands.

The United States Mining Laws

The Act of May 10, 1872, is the basic mining law governing the appropriation and purchase of mineral lands on the public domain. The Act of February 25, 1920, which removed the so-called leasable minerals from the operation of the mining laws, was the most significant mining legislation since the 1872 act. Only minor legislation was enacted between 1920 and the early 1950's.

The Nation's expanding economy and population created an increasing demand for public lands and resources, which in turn caused land use conflicts. Beginning in 1953, Congress, in the interest of resolving such conflicts, enacted legislation which significantly altered the mining laws. These acts were passed in 1953, 1954, and 1955. They are commonly referred to as the multiple use mining laws.

Public Law 250 (Act of August 12, 1953, 67 Stat. 539) and Public Law 585 (Act of August 13, 1954, 68 Stat. 708) permit the multiple development of both leasable and locatable mineral deposits on the same tract of land under the mining and mineral leasing laws. Claims for locatable minerals which were validated pursuant to Public Law 250 and Public Law 585 or located after August 13, 1954, must contain a reservation to the United States of the leasable minerals upon going to patent, if the lands were included in an application, permit, or lease or were known to be valuable for leasable minerals.

Public Law 167 (Act of July 23, 1955, 69 Stat. 367) provides for the multiple use of the land and surface resources on mining claims. It also removes common varieties of sand, gravel, cinders, pumice, pumicite, and clay from the purview of the mining laws and places them under the Materials Act of July 31, 1947.

Public Law 357 (Act of August 11, 1955, 69 Stat. 679) permits mining claims to be staked for uranium on lands classified or known to be valuable for coal.

Public Law 359 (Act of August 11, 1955, 69 Stat. 681) restored to mining location approximately 7 million acres of land withdrawn or reserved for power development.

In the closing days of the 85th Congress two new acts significantly amended the mining laws. Public Law 736 (Act of August 23, 1958, 72 Stat. 829) changed the period for doing assessment work so that each year for assessment work purposes will begin on September 1 instead of July 1.

Public Law 876 (Act of September 2, 1958, 72 Stat. 1701) provides that geological, geochemical, and geophysical surveys may be used for the fulfillment of annual labor assessment requirements. Such surveys may fulfill the requirements for two consecutive years, but may not count toward the $500 of improvements on a mining claim necessary to obtain patent.

Meeting the Public

In your contacts with the public, you will be asked to give advice or an opinion on many problems. When this occurs you should remember that you are not an attorney nor are you acting as a professional consultant. Therefore, on legal questions confine yourself to reading the law and referring to pertinent decisions. Refer the person to his attorney for legal advice. When the problem is of a technical nature, guard against giving information, advice, or suggestions which are not within BLM's jurisdiction or authority.

Frequently examiners are asked questions by public land claimants whose cases are being investigated. In these instances, limit answers to those of a general nature, or to an explanation of procedures involved. This is especially true in talking to mineral claimants who have claims on lands under the jurisdiction of other Government agencies. The mineral examiner's job is to secure facts, not to adjudicate rights. The importance of insuring the public's understanding of the Bureau and its functions cannot be overemphasized. Publicity of BLM mineral activities in newspapers, mining publications, and by other communication media is a suitable way to promote good public relations.

Often the only contact a mining claimant has with the Bureau is through a mineral examiner. The impression which is left with the claimant will depend largely upon the manner in which the engineer makes his contact and conducts himself during the examination. The examiner should exhibit a friendly and courteous attitude at all times. The claimant should be invited to accompany the examiner and to indicate the places from which he wishes the samples to be taken, or at least be consulted regarding sample points selected by the engineer. This does not prevent the engineer from taking samples from other locations not selected by the claimant necessary to perform an adequate professional examination of the mining claim. On the other hand, the examiner should not discuss his opinion on the case or imply his conclusions.

Authority To Examine Mining Claims

The general authority of the Secretary of the Interior with respect to public lands is set forth in *Cameron* v. *United States*, 252 U.S. 450 (1920) where the court said:

> By general statutory provisions the execution of the laws regulating the acquisition of rights in the public lands and the general care of these lands is confided to the Land Department, as a special tribunal; and the Secretary of the Interior, as the head of the Department, is charged with seeing that this authority is rightly exercised to the end that valid claims may be recognized, invalid ones eliminated, and the rights of the public preserved. . . . [cases cited]
>
> . . . the power of the Department to inquire into the extent and validity of the rights claimed against the Government does not cease until the legal title has passed.
>
> . . .
>
> [The Department's] province is that of determining questions of fact and right under the Public Land Laws, of recognizing or disapproving claims according to their merits, and of granting or refusing patents as the law may give sanction for the one or the other. . . .

The Department has the power to determine questions of fact under the public land laws. A BLM mineral examiner who has been properly authorized may go on an unpatented mining claim to investigate it. Every effort should be made for amicable entry and examination of the claims, preferably accompanied by the mining claimant or his duly appointed representative.

If the mining claimant threatens or uses force to prevent the examiner from going on the land, the State supervisor should be notified. If he is unable to get the claimant to agree to the examination, the U.S. Marshal may be contacted to give protection. In some instances it may be necessary to work through the U.S. Attorney to secure the Marshal's participation.

Professional Ethics

The examiner should conduct himself at all times in a manner befitting his profession. As Federal employees, the Code of Ethics for Government Service is also binding on mineral examiners in carrying out their assigned duties.

The investigation and examination should be conducted with an open and impartial mind. The examiner should be clear, concise, and factual, taking special care to avoid statements which do not clearly distinguish between fact and opinion.

The examiner's professional reputation is at stake every time he conducts an examination. At a hearing, he is strictly "on his own." All opinions, conclusions, and recommendations should be firmly backed by facts and information obtained during the investigation.

The Canons of Ethics for Engineers is a guide to professional conduct. Not all sections are entirely applicable to Government engineers, but in general the code should be followed in the performance of professional duties.

CANONS OF ETHICS FOR ENGINEERS

Foreword

Honesty, justice, and courtesy form a moral philosophy which associated with mutual interest among men, constitute the foundation of ethics. The engineer should recognize such a standard, not in passive observance, but as a set of dynamic principles guiding his conduct and way of life. It is his duty to practice his profession according to these Canons of Ethics.

As the keystone of professional conduct is integrity, the engineer will discharge his duties with fidelity to the public, his employers, and clients, and with fairness and impartiality to all. It is his duty to interest himself in public welfare, and to be ready to apply his special knowledge for the benefit of mankind. He should uphold the honor and dignity of his profession and also avoid association with any enterprise of questionable character. In his dealings with fellow engineers he should be fair and tolerant.

Professional Life

Sec. 1. The engineer will cooperate in extending the effectiveness of the engineering profession by interchanging information and experience with other engineers and students and by contributing to the work of engineering societies, schools, and the scientific and engineering press.

Sec. 2. He will not advertise his work on merit in a self-laudatory manner, and he will avoid all conduct or practice likely to discredit or do injury to the dignity and honor of his profession.

Relations With the Public

SEC. 3.   The engineer will endeavor to extend public knowledge of engineering, and will discourage the spreading of untrue, unfair, and exaggerated statements regarding engineering.

SEC. 4. He will have due regard for the safety of life and health of the public and employees who may be affected by the work for which he is responsible.

SEC. 5. He will express an opinion only when it is founded on adequate knowledge and honest conviction while he is serving as a witness before a court, commission, or other tribunal.

SEC. 6. He will not issue ex parte statements, criticisms, or arguments on matters connected with public policy which are inspired or paid for by private interests, unless he indicates on whose behalf he is making the statement.

SEC. 7. He will refrain from expressing publicly an opinion on an engineering subject unless he is informed as to the facts relating thereto.

Relations With Clients and Employers

SEC. 8.   The engineer will act in professional matters for each client or employer as a faithful agent or trustee.

SEC. 9.   He will act with fairness and justice between his client or employer and the contractor when dealing with contracts.

SEC. 10.   He will make his status clear to his client or employer before undertaking an engagement if he may be called upon to decide on the use of inventions, apparatus, or any other thing in which he may have a financial interest.

SEC. 11.   He will guard against conditions that are dangerous or threatening to life, limb, or property on work for which he is responsible, or if he is not responsible, will promptly call such conditions to the attention of those who are responsible.

SEC. 12.   He will present clearly the consequences to be expected from deviations proposed if his engineering judgment is overruled by nontechnical authority in cases where he is responsible for the technical adequacy of engineering work.

SEC. 13.   He will engage, or advise his client or employer to engage, and he will cooperate with, other experts and specialists whenever the client's or employer's interests are best served by such service.

SEC. 14.   He will disclose no information concerning the business affairs or technical processes of clients or employers without their consent.

SEC. 15.   He will not accept compensation, financial or otherwise, from more than one interested party for the same service, or for services pertaining to the same work, without the consent of all interested parties.

SEC. 16.   He will not accept commissions or allowances, directly or indirectly, from contractors or other parties dealing with his client or employer in connection with work for which he is responsible.

SEC. 17.   He will not be financially interested in the bids as or of a contractor on competitive work for which he is employed as an engineer unless he has the consent of his client or employer.

SEC. 18.   He will promptly disclose to his client or employer any interest in a business which may compete with or affect the business of his client or employer.   He will not allow an interest in any business to affect his decision regarding engineering work for which he is employed, or which he may be called upon to perform.

Relations With Engineers

SEC. 19.   The engineer will endeavor to protect the engineering profession collectively and individually from misrepresentation and misunderstanding.

SEC. 20.   He will take care that credit for engineering work is given to those to whom credit is properly due.

SEC. 21.   He will uphold the principle of appropriate and adequate compensation for those engaged in engineering work, including those in subordinate capacities, as being in the public interest and maintaining the standards of the profession.

SEC. 22.   He will endeavor to provide opportunity for the professional development and advancement of engineers in his employ.

SEC. 23.   He will not directly or indirectly injure the professional reputation, prospects, or practice of another engineer. However, if he considers that an engineer is guilty of unethical, illegal, or unfair practice, he will present the information to the proper authority for action.

SEC. 24.   He will exercise due restraint in criticizing another engineer's work in public, recognizing the fact that the engineering societies and the engineering press provide the proper forum for technical discussions and criticism.

SEC. 25.   He will not try to supplant another engineer in a particular employment after becoming aware that definite steps have been taken toward the other's employment.

SEC. 26.   He will not compete with another engineer on the basis of charges for work by underbidding, through reducing his normal fees after having been informed of the charges named by the other.

SEC. 27.   He will not use the advantages of a salaried position to compete unfairly with another engineer.

SEC. 28.   He will not become associated in responsibility for work with engineers who do not conform to ethical practices.

Adopted by Engineers' Council for Professional Development,
October 25, 1947

CHAPTER 2

SAFETY

General Hazards

In making mineral examinations the engineer will often have to work alone. He may be miles from the nearest source of help, generally in mountainous and rugged terrain. In case of mishap he must depend entirely on his own knowledge and resourcefulness. Proper preparation prior to going into the field and the use of sound safety practices while in the field will reduce the chances of mishap to a minimum. Before traveling or working in a new area, the engineer should determine the hazards to be met and learn what precautions are needed to overcome them. The following check list will guide in preparing for field trips:

1. Check all equipment, including vehicles, to make sure each item is in proper operating condition.

2. Be sure to have adequate map coverage for the area to be visited. Obtain all pertinent data on the condition of roads and trails and possible hazards, for example, sudden ground blizzards, flash floods, fire conditions.

3. Always file an itinerary with the office prior to going into the field. Notify the office of major change so that you may be located if necessary.

4. Always carry a first aid kit, and when desirable, a snake bite kit, with field equipment.

5. On any trip into remote areas, emergency food and water should be included as well as matches and bedroll or sleeping bag.

6. Safety glasses or goggles should be a standard item of field equipment.

7. Each employee who operates a motor vehicle must have a motor vehicle operator indentification card and pass a driving test. Traffic violations, preventable accidents, and other misuse of the vehicles may result in suspension or revocation of this identification card and may seriously hamper your career with the Bureau.

8. Vehicles shall be provided with emergency equipment appropriate to the terrain, climate, and weather conditions, such as tire chains, shovel, ax or pulaski, tow chain, and handyman jack. In addition, tools for emergency mechanical repairs should be carried.

9. All accidents and injuries should be properly reported to the examiner's immediate supervisor. If he cannot be contacted quickly report injuries and accidents to the nearest Bureau office.

10. Additional information on safety may be found in the Bureau Safety Handbook.

Special Hazards of Mine Workings

An underground mine is a hazardous place and should never be entered alone. No engineer should enter any mine which good judgment indicates is unsafe. No engineer should enter a mine without the protection of a hard hat and an adequate light source. As a safety precaution, an auxiliary light and a candle to check the oxygen should be carried. A mine safety lamp is preferred.

The back or roof of the working should be constantly watched for loose material that can fall if unsupported. If the mine is timbered and lagged, beware of rotted timber. Watch for winzes hidden by water in wet workings or by rotted timber in other working.

Caution must be used in checking for bad air, air lacking in sufficient oxygen, air containing poisonous gases or both. Hydrogen sulfide can be detected by its odor of rotten eggs. The following chart shows the relative dangers of gases that the engineer is most likely to encounter. Study the Bureau of Mines Circular 33, *Mine Gases and Methods for Detecting Them* (revised March 1954), by J. J. Forbes and G. W. Grove.

Mine Gases and Their Hazards

	Normal percent	Candle extinguishes	Breathing difficult	Carbide lamp extinguishes	Death danger	Dangerous to breathe
Oxygen, O_2	21.0	16.0	15.0	13.0	10.0	No.
Carbon dioxide, CO_2	0.03	--------	5.0–6.0	--------	18.0	%.
Carbon monoxide, CO	--------	--------	--------	--------	0.05	Yes.
Methane, CH_4	Explosive (From 5 to 15%; maximum at 9%)				--------	No.
Hydrogen sulfide, H_2S	--------	--------	--------	--------	0.1	Yes.
Nitrous oxide, N_2O	All other oxides of nitrogen are toxic				--------	No.
Nitrogen, N_2	78.0	--------	--------	--------	--------	No.

CHAPTER 3

TYPES OF EXAMINATION

The Mineral Examiner's Task

As a team member of the BLM activity which is responsible for the administration of the mining laws, a mineral examiner's role is a varied and important one. A mineral examiner must have thorough knowledge of the public laws which govern mineral resource management and development. An understanding of the many decisions of the Bureau, the Department, and the courts is also desirable.

A mining claimant who has complied with the law has received a possessory interest in Federal lands which at his option can be converted into a fee title. The mineral examiner must determine the claimant's compliance with the law by examining the claim, by interviewing operators or other persons engaged in the mining industry, and by reviewing technical publications on the area or subject of the examination. The mineral examiner should examine the evidence to find facts and indications which prove or disprove the mineral character of the land, the extent of minerals, and the extent of efforts to develop the lands for their mineral content.

With the facts at hand, it is the mineral examiner's function as an expert to apply the standards recognized by law and to give an opinion as to whether the mining claims under examination meet those standards. If it is the examiner's conclusion that proper standards have not been met, and if that conclusion is concurred in by the reviewing official in the Bureau, contest action may be recommended. Where charges are answered and a hearing is held, the examiner will be required to testify as an expert witness for the Government.

Mineral Patent Applications

The Federal regulations pertaining to locating and patenting mineral lands on the public domain is in 43 CFR, Part 185. Part 185.38 through 185.72 deals with the regulations governing the patenting of lode claims, mill sites, and placer claims.

A copy of the mineral patent application is received by the minerals office from the land office after the status has been checked and pre-adjudication completed. A copy of the mineral survey plat should be obtained from the survey office and the survey notes checked for pertinent data.

To aid location in the field, the claim may be sketched on a working map together with ties to section or quarter corners or other prominent landmarks. The claimant should then be advised of a proposed date of examination and invited to accompany the mineral examiner in order to indicate the points of discovery, improvements claimed, and other information pertinent to the application for patent.

The mineral examiner will examine each claim to recommend whether or not a valid discovery has been made as required under 43 CFR, Parts 185.3 and 185.12. The Department has followed the "prudent man" rule as set forth in *Castle* v. *Womble*, 19 L.D. 455, in determining the validity of a discovery. That rule holds a claim valid where mineral is found in such quantity and of such quality as would warrant a prudent man in the expenditure of his labor and means in an effort to develop a paying mine.

The mineral examiner must exercise good judgment and knowledge in developing all the data pertinent to a discovery. On the basis of such data—the actual mineral showing and its relation to the geology of the area and in the knowledge of its significance with respect to the discovery of ore deposits in the district, the mineral examiner must decide if there is a sufficient showing of mineral to validate a discovery under the "prudent man" rule. He must distinguish between a showing sufficient to validate a discovery and one only sufficient to warrant further prospecting or exploration in the hope of making a discovery. Geological inference, no matter how strong or convincing, cannot be used as a discovery in lieu of a showing of mineral in place.

Decisions by the Director and the Secretary have held that geophysical or geochemical data will not qualify as mineral discoveries without visual exposure of the minerals claimed. Discoveries of mineral showings claimed to have been made in drill holes may be accepted under certain conditions. Such conditions are visual inspection of drill sites, availability of well kept drill logs, verifiable data on drilling contracts.

While the regulations for obtaining a mineral patent provide that the point of discovery must be designated on the mineral survey plat, it is not required that the actual discovery be made in the "discovery" working. A discovery of mineral that can qualify under the "prudent man" rule must be made within the boundaries of each claim, but it does not have to be on or near the surface or at any specific point.

Classification

Classification of the public domain, acquired lands, and certain private lands is an important part of the mineral examiner's work. The classification of land as mineral or nonmineral in character must be based on facts developed from a careful study of the pertinent literature and a careful study of the land in question. Geologic inference is of primary importance and should be used in determining whether or not the land should be classified mineral in character. (See *United States* v. *Standard Oil Company of California*, 20 Fed. Supp. 432–63 Public Lands—1937, which states:

> For determination of the question whether lands can be declared by the land Department to be known mineral, any competent, relevant evidence, any deduction by men skilled in the field, based on observable or deducible facts or indicia on the land or in its vicinity or on deductions from the geological formation of the area, disclosures or other surrounding or external circumstances, can be relied on.

The authority and necessity for mineral classification to effect title transfer of public lands is shown in the following chart.

Type of disposition	Authority	Title 43 CFR reference	Effect of existence of locatable minerals [1] [2]
Homestead (original).	Homestead (original)—Act of May 20, 1862 (12 Stat. 392) as amended.	Part 166___ Part 102.	Not suitable for entry if mineral in character, 43 CFR 166.2.
Homestead (enlarged).	Homestead (enlarged), Act of Feb. 19, 1909 (35 Stat. 639; 43 U.S.C. 218) as extended and Act of June 17, 1910 (36 Stat. 531, 43 U.S.C. 219).	Part 167___ Part 102.	Not suitable for entry if mineral in character, 43 CFR 167.1c.
Desert land entries.	Act of March 3, 1877 (43 U.S.C. 321–323) amended by Act of March 3, 1891 (43 U.S.C. 321, 323, 325, 327–329).	Part 232___ Part 102.	Not suitable for entry if mineral in character, 43 CFR 232.3.
Selections] (State).	Act of Feb. 28, 1891 (26 Stat. 796; 43 U.S.C. 851).	Part 270___ Part 102.	Not suitable for selection if mineral in character, 43 CFR 270.1, unless base lands are also mineral in character (72 Stat. 938).—See Director's memo. 5.04e:D of 9–16–58. Alaska lands exception—P.L. 85–508 (76 Stat. 339).
Exchanges (private).	Act of June 28, 1934 (48 Stat. 1272) as amended.	Part 146___ Part 102.	Either party can make reservations—43 CFR 146.1—reservations in Vol. V 2.15.2 BLM Manual. Mineral values should be included in appraisal, if they are to be transferred.
Exchanges (States).	Act of June 28, 1934 (48 Stat. 1272) as amended by Act of June 26, 1936 (43 U.S.C. 315g).	Part 147___ Part 102.	Minerals may be reserved by either party. Mineral values should be included in appraisal if they are to be transferred, 43 CFR 147.2.
Public sales_____	Section 2455 of Rev. Statutes as amended by Sec. 14 of Act of June 28, 1934 (43 U.S.C. 1171) and Act of July 30, 1947 (61 Stat. 630).	Part 250___ Part 102.	Not suitable for sale if mineral in character; 43 CFR 250.8.

See footnotes at end of table.

Type of disposition	Authority	Title 43 CFR reference	Effect of existence of locatable minerals [1] [2]
Small tracts (sale or lease).	Act of June 1, 1938 (52 Stat. 609) as amended by Act of June 8, 1954 (68 Stat. 239).	Part 257... Part 102.	All minerals are reserved in patent. Leasable minerals may be leased. No provision for prospecting or disposition of locatable minerals. Land must be chiefly valuable for purpose classified, so could be classified as chiefly valuable for locatable minerals, if circumstances warrant such action. 43 CFR 257.1 and 257.16.
Recreation and Public Purposes Act (sale).	Act of June 14, 1926 (44 Stat. 741), as amended by the Act of June 4, 1954 (68 Stat. 173).	Part 254... Part 102.	All minerals are reserved. Leasable minerals may be leased. No provision for prospecting or disposition of locatable minerals. 43 CFR 254.14. Land must not be more valuable for some other use, such as mineral development, 43 CFR 254.5a.
Color of title (general).	Act of December 22, 1928 (45 Stat. 1069) as amended by the Act of July 28, 1953 (67 Stat. 227).	Part 140... Part 102.	Minerals may be reserved or included in patent, depending on characteristics of color of title claim and existence of mineral permits. No question regarding existence of locatable minerals, 43 CFR 140.8.

[1] There are no specific statutes relating to the reservation of *saleable* minerals. Thus the treatment would be the same as for locatable minerals.

[2] Leasable minerals may be reserved and the land classified for disposition unless the lands are classified as not suitable for disposition by the Geological Survey. (See BLM Manual Vol. 6—Minerals, Chapter 5.14.)

Conflicts

Mineral entries or locations on the public domain often conflict with applications for other types of acquisition such as small tracts, homesteads, desert land entry and material sites; with other uses such as range improvement projects, timber sales, and material sales; and with various withdrawals and rights-of-way.

The existence of mining claims on the public domain is considered *prima facie* evidence of the mineral character of the land until proven otherwise, particularly if the claims were located prior to initiation of other interests as described above and prior to the act of July 23, 1955. The existence of a conflict usually becomes evident when the land is examined for classification as to its mineral character or suitability for homestead, small tract or other uses mentioned above. When such a conflict is discovered, it is necessary to determine the validity of the mining claim before disposition of the nonmineral entry is made.

Although conflicts may exist between a mining claimant and a leasable mineral lessee, the Bureau is not ordinarily concerned with private contests resulting from *in rem* proceedings brought pursuant to Public Law 585. The accompanying chart shows the classification of mining claims in regard to the date of location under the act of August 13, 1954.

Mining claims located on vacant lands later covered by mineral leases or permits before August 13, 1954.	Valid.
Mining claims located before July 31, 1939, on lands covered by mineral leases or permits or applications for same or known to be valuable for leasable minerals.	Invalid.
Mining claims located between July 31, 1939, and February 10, 1954, on lands covered by mineral leases or permits or applications for same or known to be valuable for leasable minerals.	Valid only if compliance had with provisions of PL 250 and/or PL 585.
Mining claims located between February 10, 1954, and August 13, 1954, on lands covered by mineral leases or permits or applications for same or known to be valuable for leasable minerals.	Invalid.
Mining claims located after August 13, 1954, on lands covered by mineral leases or permits or applications for same or known to be valuable for leasable minerals.	No conflict because of separation of rights to locatable and leasable minerals.

Public Law 167

When range improvement projects or timber sales are to be programmed, it is desirable that the Bureau of Land Management have the right to manage and sell the surface resources. The act of July 23, 1955, Public Law 167, provides for a procedure whereby the Government may obtain surface management rights on claims located prior to the act. Claims located after July 23, 1955, are subject to the provisions and limitations of the act, including the Government's right to manage the surface resources.

Claims included in a verified statement which is filed in response to a notice to mining claimants must be examined for a valid discovery. Proceedings are initiated on those claims that do not appear to have a valid discovery in order to determine, at a hearing, whether the claimant or the Government has the right to manage the surface resources. However, the claimant may file a withdrawal, waiver, or stipulation for claims covered by his verified statements, in which case no hearing is held. Those claims determined to have a valid discovery are not subject to the provisions of the act.

Even though the Government may gain the right to manage the surface and surface resources as the result of Public Law 167 proceedings, the mining claimant does not lose any possessory rights to the minerals or his rights to mine and to use as much of the surface as is necessary to his operations. Further, any permittee or licensee of the Government may not endanger or materially interfere with prospecting, mining, or processing operations or uses reasonably incident thereto.

If it is in the interest of the Government to exercise exclusive jurisdiction over the land and the evidence indicates that the mining claims involved are invalid, a regular contest action should be brought. If, however, only the use of the surface and use and disposition of the vegetative surface resources are desired and if mining operations would not unreasonably interfere with such uses, then Section 5 of the act should be utilized. Also, more than one proceeding against the same mining claims or claimants within a relatively short period of time should be avoided. Thus, in the absence of unusual circumstances, a proceeding should not be initiated under the act with respect to mining claims in a given area, if the Bureau knows or has reason to believe that it will have cause to bring adverse proceedings (contest actions) against such claims.

Property Valuation

Valid mining claims, located in areas of the public domain which are later required by the Government for Federal highway rights-of-way, various reclamation projects, and other uses, must be appraised for purchase or condemnation procedures. Such appraisals or valuations are part of the job of the valuation engineer. In addition to appraising mining claims and other mining property, the valution engineer is often required to appraise mineral deposits such as sand and gravel, stone, pumice, or cinders for material sales.

Reimbursable Cases

Frequently the Bureau of Land Management is called upon by other governmental agencies to determine the presence and validity of mining claims. Authority for conducting these investigations is usually a "Memorandum of Understanding" outlining the work to be done, the estimated cost, and providing for reimbursement.

Special Cases

Special cases include the examination of mining claims under the provisions of Public Law 585, 359, and 357; mineral and occupancy trespasses involving mining claims and mill sites; and preliminary examinations under Public Law 167. These examinations may or may not require validity determinations. They generally involve considerable status and record checking to see if the laws and regulations have been complied with.

CHAPTER 4

EXAMINATION PROCEDURES

Preliminary Preparation

After assignment of the case, the mineral examiner should prepare for the field examination.

He should carefully examine the case file, making certain that it contains all available pertinent case history and that the status is complete and adequate. If additional status is required, submit a request to the Land Office detailing the status wanted.

He should obtain a suitable work map and plot all necessary available data on it. The official copies of the plat should never be used as a work copy. Sometimes it is desirable to use more than one map, depending upon the maps available and the work to be done. Some of the best sources outside the Bureau for maps are:

a. The Geological Survey,
b. The Forest Service, U.S. Department of Agriculture,
c. State highway departments and county road maps,
d. The Army Map Service,
e. The Atomic Energy Commission, and
f. Aeronautical navigation charts.

In addition to maps, aerial photographs are very useful and highly desirable for some cases, particularly where area geology is important. When possible, they should be obtained in stereo pairs.

Available literature should be reviewed for articles, reports, papers, etc., covering the area involved in the examination. Pertinent data added to the working map is often quite helpful. Readily available sources of information are as follows:

a. Geological Survey,
b. Bureau of Mines,
c. State Bureau of Mines and Geology,
d. State university and mining schools,
e. Journals on mining, geology, metallurgy, and petroleum, and
f. Publications of the various mining, geological, and petroleum societies.

Equipment should be carefully selected and checked to insure that all items required are on hand and in operating condition. Prior to going out, set Brunton compass for the proper declination. For proper care of samples make sure that clean, uncontaminated sample bags are used. Be sure that plenty of sample cutting tools are carried, especially if sharpening equipment cannot be taken into the field. Tungsten carbide tipped moils are recommended for cutting channel samples. Be sure that an adequate supply of sample tags, Form 4–1448 (July 1959), are taken on all examinations.

The mineral examiner should be equipped and prepared to take pictures on all field assignments. The right photographs properly presented can be the most convincing testimony or evidence offered. For this reason the engineer should always use a good panchromatic or all color sensitive film for black and white prints. Occasionally, color pictures may be desirable, and at times pictures will have to be taken underground. It is recommended that a good 35mm camera with flash attachments and tripod be used, as it is most adaptable to our work.

A suggested checklist of items which should be available for making field examinations follows:

1. File containing copies of material pertinent to the field exam.

2. Maps and area data.

3. Official field notebook with ample supply of sample tags and notepaper.

4. Clear plastic protractor scales for 10–20, 20–40, and 30–60 scales.

5. Pencils, including color set.

6. Brunton compass.

7. Pocket tape (8 ft.—inches and tenths) and 50 ft. or 100 ft. engineer's metallic or steel tape.

8. Camera, attachments, and film.

9. Clip board.

10. Hand lens, 10X.

11. Geologist's pick.

12. Clean sample sacks (appropriate sizes) and sample bottles.

13. Hammer, chisels, and moils.

14. Safety goggles.

15. Hard hat.

16. Carbide lamp, safety lamp or candles, flashlight, or lantern.

17. Geiger counter or scintillometer.

18. Mineral light.

19. Alnico magnet.

20. Gold pan.

21. Roundpoint shovel and pick (miner's or railroad).

22. Field glasses or binoculars.

Field Examination

The first step in making an examination is to locate the boundaries of the area to be examined. A reconnaissance of the area is then made, noting the dominant features which pertain to or have a bearing on the examination. These features would primarily be concerned with the geology and topography, and to a somewhat lesser extent, with vegetation, trails, roads, water, and power sources. In so far as possible, these features should be noted on the work map while the reconnaissance examination is being made.

The purpose of the reconnaissance is to acquaint the examiner with the area and provide him with information which will enable him to plan his approach to the issue and to proceed with the examination in the most efficient manner. In validity determinations, the issue is to determine whether or not the claimant has developed a sufficient showing of mineral in place to constitute a discovery under the "prudent man rule" which is set forth in *Castle* v. *Womble*, 19 L.D. 455. *In all validity examinations, the claimant should be notified and given an opportunity to be present during the examination.*

In examining a mining claim to determine whether or not a valid discovery has been made, the mineral examiner should bear in mind that it is not his job to make a discovery, but merely to verify one. The discovery points should be available and safe for examination. If the alleged discovery is in a shaft or underground working which is not accessible or safe to enter, the engineer should not try to make the discovery accessible or enter under unsafe conditions. The mining claimant should be requested to make the discovery points accessible. If he refuses, the claim may be contested on the basis of evidence available and the validity determined at a hearing.

When the discovery point is accessible, it should be carefully examined and mapped, showing all structures and their attitudes and points sampled. An adequate number of samples should be taken to be representative of the tenor of the mineral deposit.

Lode Claims

For lode claims, surface structures should be plotted on the map of the claim and correlated with any structure shown in the discovery shaft, cuts, workings, and so forth. Additional information should include a description of the ore minerals, gangue minerals, vein and wall-rock alteration, and the country rock. Show manner in which mineral occurrence is or is not similar to other successful discoveries or operating mines in the district. A detailed description of all workings and improvements on each claim should be made. When deemed necessary, workings should be plotted on the work map.

Placer Claims

In general, all forms of valuable mineral deposits which are not in veins or lodes are locatable as placers, usually conforming to the public land survey. There are two general types of placer mineral deposits, (1) mineral particles, either metallic or nonmetallic, of relative high specific gravity, disseminated in an unconsolidated ground mass, and (2) massive deposits of nonmetallic, low unit value mineral substances of nearly uniform physical or chemical composition.

To constitute a valid discovery, the minerals must be present in sufficient quantity and quality to justify a prudent man in the further expenditure of his time and means in an effort to develop a paying mine. Further, as a criteria applied to the second type of placer deposits, it must be shown that the mineral deposit is valuable by virtue of being mined and marketed at a profit.

The Act of July 23, 1955, P.L. 167, removed common varieties of sand, stone, gravel, pumice, pumicite, cinders, and clay from location under the mining laws. Therefore the claimants of placer claims located for such mineral materials prior to the act must be able to show that at the time the act was passed the deposits were valuable, in demand, and could be produced and marketed at a profit. After July 23, 1955, only mineral materials of "uncommon" variety and possessing special values may be located under the mining laws.

The accompanying chart shows the effect of Public Law 167 on the disposition of sand and gravel from unpatented mining claims.

Chart To Accompany Solicitor's Opinion M–36467, August 28, 1957, "Disposal of Sand and Gravel From Unpatented Mining Claims"

A. *Unpatented Placer Mining Claims Located Before July 23, 1955 (P.L. 167)*

(1) Sand and gravel was discovered on claim and at date of location was a valuable mineral; it occurs with or without other valuable minerals (gold, etc.) on the claim.

Claim is valid and owner may sell his sand and gravel; U.S. may *not* sell this sand and gravel.

(2) Sand and gravel is *not* a valuable mineral; it occurs with valuable minerals (gold, etc.) on the claim; claim has a discovery based on such valuable minerals, in accordance with the prudent man conception.

Claim is valid, but owner may *not* dispose of sand and gravel as a by-product; he may use it to the extent that it can be utilized in the mining and removal of valuable minerals from the claim, U.S. may *not* sell this sand and gravel.

(3) Sand and gravel is *not* a valuable mineral; it occurs without a valuable mineral (gold, etc.) on the claim; there is no discovery on the claim in accordance with prudent man conception.

Claim is invalid. If declared null and void and cancelled United States may sell sand and gravel.

B. *Unpatented Placer Mining Claims Located on or After July 23, 1955 (P.L. 167)*

(4) Sand and gravel of "common variety" occurs on the claim; it may or may not be a valuable mineral; it occurs with valuable minerals (gold, etc.) that are not a "common variety"; there is a discovery on the claim in accordance with prudent man conception, based on one or more valuable minerals that are not a "common variety".

Claim is valid, but owner may *not* sell sand and gravel, he may use it for any mining purposes; U.S. may *not* sell this sand and gravel.

(5) Sand and gravel of "common variety" occurs on the claim; it may or may not be a valuable mineral; it does not occur with valuable minerals (gold, etc.); there is *no* discovery on the claim in accordance with the prudent man conception and based on one or more valuable minerals that are not a "common variety".

Claim is invalid. After final determination of invalidity of claim U.S. may sell this sand and gravel. Owner may *not* sell sand and gravel off this claim. The taking of sand and gravel from a mining location by one with knowledge of the existence and date of location of the claim, is a willful trespass.

Although only one discovery is required for each placer claim, whether of 20 acres by an individual locator or of 160 acres by an association of eight persons, the mineral character of each 10-acre parcel, usually in square form, must be verified by the mineral examiner. As in the examination of lode claims, the character of the deposit should be fully described. Discovery points, workings, improvements, and roads should be sketched on a work map.

Where marketability is at issue, the investigation properly includes information on it. Interviews with operators, production cost data, market conditions (present and prospective), transportation costs, and other pertinent data should be developed in this phase of the examination.

When association placers are involved, any indication of dummy locators or other fraudulent actions suspected by the examiner should be carefully and fully documented in the field notes for further consideration in the case. Especially in patent application cases, irregularities may be noted by the adjudicator, but the field examiner will have the task of investigating and reporting on them.

Patent Applications

In addition to examining mining claims included in a patent application for compliance with discovery requirements mentioned under lode and placer claims, the engineer should be careful to note that the corners of the claims are properly marked, that notice of application and plat were properly posted in a conspicuous place, and that a reasonable estimate of the value of improvements for each claim totals at least $500. If more than one claim is involved in the application, the $500 expenditure need not have been made on each claim independently but may be expended on one or several of a group, provided that the total expenditure is equivalent to $500 per claim, or more, and that the improvements made clearly benefit and tend to develop the group as a whole.

Mineral Classification

The object of making a mineral classification is to determine whether or not the land in question is valuable or prospectively valuable for mineral deposits at the time the classification is made.

In contrast to validity determinations, the use of geologic inference in making a mineral classification may be of primary importance. The determination may be made on the basis of information obtained from the literature in conjunction with a careful study of the land. During the examination, all evidence of a geologic nature will be noted, including a classification of the general country rock, with dip and strike of beds, if they are sediments; structural features, if they are metamorphic; or textural features, if they are igneous. Note dip, strike, and trend of faults and dikes or other intrusions and the general pattern of any prominent joint systems.

Any evidence of mineralization should be carefully described and sampled, if necessary, to verify mineral values. Any correlation between faults, or faults and intrusions, indicating possible intersections favorable to ore deposition should be shown. Workings and discovery points on all mining claims should be examined for mineralization and geologic evidence. All evidence, either positive or negative, bearing on the mineral character of the land should be noted. Any photographs showing pertinent evidence should be taken and used in the classification report. The notes should also include a description of the topography, vegetation, water resources, roads, power, and so forth.

Lindley on Mines, 3rd edition, section 98 states: "The mineral character of the land is established when it is shown to have upon or within it such a substance as:

a. is recognized as mineral, according to its chemical composition, by the standard authorities on the subject; or

b. is classified as a mineral product in trade or commerce; or

c. such a substance (other than the mere surface which may be used for agricultural purposes) as possesses economic value for use in trade, manufacture, the sciences, or in the mechanical or ornamental arts.

It is demonstrated that such substance exists therein or thereon in such quantities as to render the land more valuable for the purpose of removing and marketing the substance than for any other purpose, and the removing and marketing of which will yield a profit; or it is established that such substance exists in the lands in such quantities as would justify a prudent man in expending labor and capital in the effort to obtain it."

The criteria for determining mineral character of land and making a validity determination of a mining claim are somewhat parallel in nature. The major difference is that in the former the criteria are wide in scope and requires a lesser degree of proof than in a determination of validity.

Geological inference is an acceptable tool for determining the mineral character, as pointed out in many departmental and court decisions.

The evidence bearing upon the character of the lands selected should not be restricted to mineral discovery or development upon these lands and to their geological formation but may extend to the discovery and development of minerals on adjacent lands, and to their geological formation, *Kern Oil Company* v. *Clotfelter*, 30 L.D. 583, 587 (1901).

In the case of *United States* v. *Southern Pacific Company* 251 U.S. 1 (1919) the Supreme Court gave particular note to the use of geological inference in determining the mineral character of the land when it said:

In order to establish the character of the lands, in this connection, as lands valuable for oil, it is not necessary that they shall have been demonstrated to be certainly such by wells actually drilled thereon and producing oil in paying quantities after a considerable period of pumping; it suffices if the conditions known at the time of patent, as to the geology, adjacent discoveries, and other indications upon which men prudent and experienced in such matters are shown to be accustomed to act and make large expenditures, were such as reasonably to engender the belief that the lands contained oil of such quantity as would render its extraction profitable and justify expenditures to that end.

The Department has said, "The determination of the mineral character of land is not dependent upon local economic and industrial conditions, subject to change in character with the ephemeral shifts in economic supply and demand for the particular deposit." *United States* v. *State of Utah*, 51 L.D. 432,

26

436 (1926) and "thus, since the appellant concedes and the record shows the land cannot be mined profitably, does not establish that the classification of the land as mineral is erroneous, *Freeman v. Summers*, 52 LD. 201, 206 (1927)."

In the Solicitor's Opinion M-36295 of August 1, 1955 (titled, "Use of special criteria to determine the mineral character of mining claims located for sand and gravel") the marketability factor is discussed at length. While this opinion concerns the factors determining a profitable market as an important element for establishing the validity of a sand and gravel mining claim, the same factors discussed are applicable to mineral character determinations.

Field Notes

The importance of adequate and legible notes cannot be overemphasized. Uniform note keeping has certain problems as no two examinations are conducted in the exact manner. Standard 7¼" x 4⅝" sheets, graph-ruled (10 x 10 is preferable) on the reverse side for sketching should be used. Mineral sample sheets (Form 4-1448) will also fit the standard 6-ring notebook. The original notes pertaining to each case shall be fastened securely and placed in the State office case file. Copies of these notes may be made by the engineer for his personal file.

A front sheet (Form 4-1548) for the notes, describing the case and purpose of the examination, should be prepared. Notes taken in the field should be recorded in sequence as the investigation progresses, usually in the order to be presented in the report. The following checklist is presented as a guide to aid in making the examination as well as to standardize note taking. It should also minimize the possibility of overlooking pertinent data that should have been recorded during the examination.

Checklist for Field Notes

1. Physical features
   a. Location: direction and distance from nearest town, mining district, county, and State.
   b. Accessibility: how reached, transportation facilities, etc.
   c. Topography: general description, elevations.
   d. Vegetation and timber, climate.
   e. Water and power facilities.
   f. Identification of land and/or mining claims.
2. Geology and mineralization
   a. General geology of lands involved (relate to area geology).
   b. Details of structure; description of veins or deposit.
   c. Mineralization: ore minerals, gangue minerals.
   d. Ore reserves: tonnage and tenor.
3. Mineral development
   a. Describe surface and underground workings, drill holes (relate to geology), prepare sketch map or verify data on existing maps.
   b. Plant and equipment: describe type, condition, and present use of buildings, improvements, and equipment.
   c. Estimated value and utility of plant and equipment.
4. Sample data
   a. Index samples from sample tags (elaborate on information included on sample tags, Form 4-1448, when necessary).
   b. Note: space is provided on the sample tags for recording assay results.
5. Other pertinent data
   a. Claim lines; monuments; discovery point on lode line, and so forth.
   b. Posting of plat and notice of patent application (when applicable).
   c. Ownership conflicts.
   d. Dummy locators.
   e. Compliance with State or local laws.
   f. Placers conforming to legal subdivision.
   g. Mineral character of 10-acre subdivisions (placer claims).
   h. Market data—for commonplace minerals.
   i. Unusual benefication or metallurgical problems.
   j. Photographs: what is shown; taken from; direction taken.
   k. Interviews.

Form 4—1548
(Sept. 1960)

**UNITED STATES
DEPARTMENT OF THE INTERIOR
BUREAU OF LAND MANAGEMENT**

MINERAL EXAMINATION FIELD NOTES

Case Number

Subject *(Patent application, etc.)*

Name of applicant(s) or claimant(s)

Claims involved

Mineral Survey No. _____

State	County

Lands involved:

Twp. _____ , Rge. _____ , Sec. _____

Date	Engineer

Figure 1

29

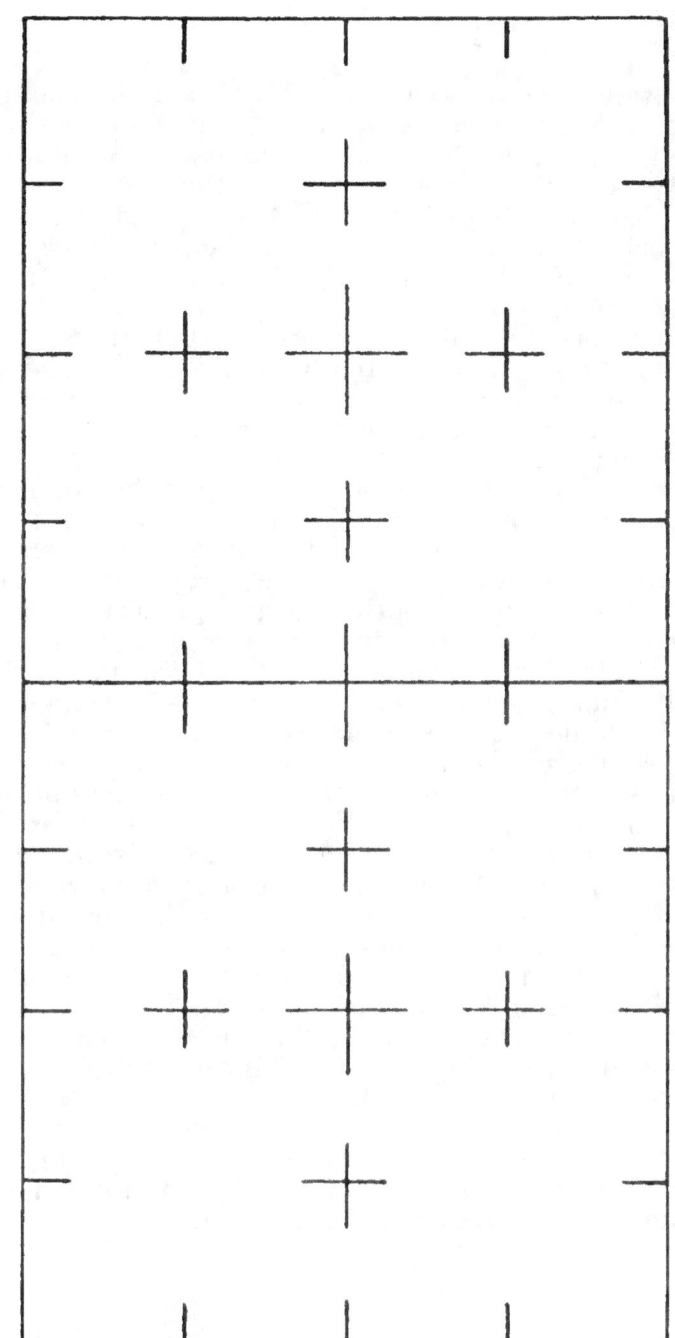

Scale: _____

Property Valuation

Property valuation differs from validity examinations in that the valuation engineer must appraise the value, considering mineral potential, of public domain, private, and State lands for exchange purposes; appraise the value of the mineral materials (including but not limited to) sand, stone, gravel, pumice, pumicite, cinders and clay for sale from the public domain upon application by the public; and appraise mineral properties in eminent domain cases.

In exchange cases it must be determined whether or not the value of the offered lands, including mineral values, is equal to or greater than that of the selected lands. The exchange will not be made unless it is to the advantage of the Government. A definite monetary figure relating to the mineral values must be given in the field reports if either or both the selected or offered lands have an actual mineral value.

In appraising mineral material deposits, the first step is to determine the composition and physical properties of the material, paying particular attention to any properties that may enhance or detract from its value. Determine how the deposit is located with respect to development, processing marketing, power, roads, and so forth. Determine if there are other similar deposits being mined in the same marketing area, and, if so, how are the deposits comparable?

Also, the engineer should try to obtain information on the uses of the material, production costs, recent comparable sales in the area, and the market value of the finished products. Determine the potential market in the area and whether the products will be restricted to the market of the immediate production area. Determine probable distance of the deposit from processing plant and market and freight facilities and costs. If there are no other deposits being worked in the area, and consequently no comparable sales, the engineer must look for the necessary data in other areas which are comparable to the one where the material is located.

In the examination and valuation of mineral properties involved in eminent domain cases, the engineer should use the following outline as a guide in making his examination, obtaining the required data, and writing his report.

Mineral Valuation Checklist

EMINENT DOMAIN CASES

Use Standard Form 4-802 for heading sheet.

A. Appraisal summary
    1. Recommended fee value
        a. Proven, probable and potential reserves. Cutoffs used.
        b. Salvage value of improvements.
        c. Net fee value.
    2. Recommended leasehold or royalty value to Government.
    3. Recommended net value per ton to Government.
    4. Recommended annual rental of lands without production.
    5. Certification of noninterest (present or future) of appraiser.
B. Purpose and scope of examination
C. Appraisal factors
    1. General
        a. Location and access-roads, trails, railroads.
        b. Topographic features in relation to property.
        c. Survey of transportation facilities, power supply, and relation of mine to markets.
        d. Summary of freight rates, power rates.
        e. Surface characteristics—vegetation.
        f. Climatic conditions: snow, rainfall.
    2. History and record status
        a. Legal
            1) Land status—include conflicts.
            2) Claim recordations — include maps showing both patented, unpatented, or leased lands.
            3) Title recorded.
            4) Changes of ownership—abstract of title.
            5) Tax history.
            6) Leases, water and mill rights, royalties, and so forth.
            7) Suits in court, claim conflicts, litigations, and so forth.
        b. Production
            1) Record of production since beginning.
            2) Record of all shutdowns and reasons.
            3) Production of surrounding district.
            4) Cooperative production from similar properties, particularly nonmetallics.
        c. Current status—producing, developing, abandoned.

d. Financial
    1) Preorganization agreements.
    2) Capitalization.
    3) Profits, dividends, and assessments.
    4) Reorganizations and consolidations.
    5) Comparison of yearly statements.
    6) Present financial structure and standing.

3. Management
    a. Personnel and organization.
    b. Character, attitude, and capability.
    c. Comparison with operators of other similar mines.
    d. Efficiency; past and present.

4. Geology
    a. Aerial geology with reference to:
        1. Geological Survey reports and maps.
        2. State publications.
        3. Publications of technical societies.
    b. Detailed geology
    Study of surface features
    Study of underground structures from:
        1. Mine maps and underground examination.
        2. Drill hole data and inspection of drill core and samples.
    c. Deductions as to formations, structures, and ore bodies.

5. Mineral improvements and mine workings
6. Sampling (Include assay plans)
    a. Study of all records.
    b. Check of previous results.
    c. Sampling of new workings.
    d. Correlation with geology.

7. Estimate of ore reserves
    a. Outlining of assured and probable ore reserves by correlation of geology, sampling, and mine maps.
    b. Tests of tonnage factor.
    c. Estimates of reserves: positive, probable and possible.

8. Methods and costs
    a. Mining
        1. Method in use; estimate of percent of recovery.
        2. Advisable improvements.
        3. Relation of development work to mining.
        4. Itemization of costs for recent periods.
        5. Estimate of most economical rate of mining and corresponding life of mine.

b. Milling
    1. Flow sheet and details of all unusual features, percent of recovery, costs.
    2. Water supply.
    3. Space available for tailings.
    4. Other factors.
9. Marketing
  a. Sale of ore or metal—show comparative marketability factors.
  b. Mill or smelter contracts.
10. Plant and equipment (include photographs)
  a. Underground
    1. Condition of shafts, drifts, cross-cuts, raises, stopes.
    2. Condition of pumps, motors, and other underground machinery.
    3. Summary of equipment at hand and additional pieces needed.
  b. Surface
    1. Condition of headframe, hoist, power plant, shops, mills, smelters, houses, hospitals.
    2. Condition of machinery.
    3. Estimate of new equipment needed.
    4. Other facilities for mining or milling.
11. Miscellaneous
  a. Timber, land, power sites, water.
  b. Labor, unions, wages, disputes, welfare, safety practices, etc. (Give prevailing rates.)
  c. Attitude of local government—smoke or dust control.
  d. Surface rights—liability for damage due to subsidence above mining operations.
D. Evaluation analysis
  1. Economic situation
    a. Average profits for recent periods.
    b. Average market in same periods.
    c. Future prospects and potential.
    d. Estimate of future earning power.
    e. Analysis of comparative marketability factors.
  2. Valuation
    a. Determination of values for years' life, profit factors, interest rates.
    b. Application of formulas and calculation of present worth.
  3. Recommended fair market value is the measure of damage (for land actually taken for public use); that is, the highest price estimated in terms of money which the land would bring if exposed for sale in the open market, with reasonable time allowed in which to find a purchaser, buying with knowledge of all the uses and purposes to which it was adapted and for which it was capable.

Sampling

The object of sampling material is to take a small portion in such a manner that the sample portion is as nearly representative of the whole as possible. In mineral examinations, the sampling is done to determine the mineral values in a deposit or discovery.

The examiner has the responsibility of the proper protection of the samples from contamination or salting from the time of sampling until turned over to a competent assay or testing laboratory. Definite plans as to the handling, splitting, and storing of the samples and the storing of the rejects should be worked out in significant detail to assure follow-up results, if necessary, on any samples obtained. This precaution may save a return trip to the property or support the examiner if questioned at the hearing.

Mineral property appraisals and validity determinations are directly tied to the results of sampling the property in question. The choice of a sampling method will generally be determined by the character of the deposit to be sampled. For example, the method of sampling will vary for placer deposits, lode or vein type deposits, and bedded deposits. Sample Form 4–1448 (July 1959) should be used to identify samples for deposits other than placer and lateritic types, and Form 4–1550 should be used for recording placer samples.

Grab samples

Grab sampling is not true sampling in the strict sense of the word even though valuable information may be obtained by it. It should only be used with care as the danger of unconsciously salting a sample is always present. Also there is always the possibility that layering or segregation of mineral values may have taken place which may not be apparent from the surface appearance and could therefore result in the sample assays being considerably different from the average of the material as a whole.

The theory of grab sampling is to gather at random a sufficiently large portion of the material to constitute a reasonable sample (at least ¼ cubic foot). This method is used only to supplement or verify results by other methods.

Sampling rock materials in place

Whether sampling a face or wall in a bedded type deposit or a vein or fissure in the face or back of a working, the area to be sampled should be thoroughly cleaned to avoid salting or dilution by material or salts that may have accumulated on the surface. If possible, a fresh surface should be developed to avoid the effects of oxidation or leaching.

The proper sampling method will vary from case to case and must, therefore, be determined by the examining engineer based upon his professional experience and judgment. The Bureau has adopted no single method of sampling, but must use all recognized methods which are standard practice in the mining industry and which sound judgment, usage, and experience have shown to be appropriate to a particular mineral deposit.

Channel samples

The usual method is to cut a uniform channel 3 to 6 inches wide and 1 or more inches deep across the section to be sampled. This method should ordinarily be used for property appraisals in eminent domain cases and for special validity cases involving properties which have been producing mines, or where there is need for special precautions.

Chip samples

By careful chip sampling, the engineer can approach the accuracy of the channel sample if the chip sample is taken progressively across the structure, either by hand pick or with a hammer and moil in amounts proportional to the quantity of material in place, being careful not to lose any material during the sampling.

Where the material to be sampled shows pronounced variations in composition due to banding or bedding, or in hardness, it is best to sample each variation separately and avoid the possibility of over-compensation for any of the variations during sampling. The amount of sample taken should, as a minimum, be about one pound per lineal foot of material sampled.

The frequency of sampling will be governed by the exigency of each case. Controlling factors include variations in composition, variation in vein width, number of veins or outcrops, number of discovery points or workings. In all cases the sample form (Form 4–1448) should be filled out completely and any contingencies not covered by the form should be explained more fully on the back of the form or in the notes with proper reference to the sample number. Sketches of the point sampled are often very useful, particularly if the case should go to a hearing or court. Unexpected or erratic assays should be rechecked by taking a new sample and carefully examining the sample points and methods.

Placer sampling

Placer sampling involves the sampling of unconsolidated or poorly consolidated materials such as sands, gravels, or alluvium containing valuable minerals or mineral materials. It differs from other types of sampling in that usually the valuable minerals are separated from the bulk of the sample in a heavy mineral concentrate at the time of sampling and reported in pounds, ounces, or grams of mineral recoverable per cubic yard. The number of samples taken will depend on the case involved. For mining claims, the samples should be related to discovery points. For classification purposes, the sampling must be adequate to determine the mineral character of the land; and for appraisal purposes, the sampling should be adequate to determine the character, properties, and value of the deposit.

Form 4—1448
(July 1959)
UNITED STATES DEPARTMENT OF THE INTERIOR
BUREAU OF LAND MANAGEMENT

_____ , State Office

MINERAL SAMPLE RECORD

Sample No. _____ Date _____

Name of Claim _____

District _____ ; Sec _____ Tp _____ Rg _____

Sample: *(Show sketch of sample point on reverse)*

Description ·

Type: ☐ Stope ☐ Dump ☐ Cut ☐ Outcrop ☐ Pit
☐ Tunnel ☐ Shaft ☐ Drift ☐ X-cut ☐ Raise ☐ Winze

Sample taken _____ ft. _____

(Distance)   (Direction)

From _____

Length _____ Width _____ Depth _____

Part of vein taken _____ Width Rep _____

Strike _____ Dip _____

Weight of sample _____ Est. moisture _____ %

Assay for ☐ Au ☐ Ag ☐ Pb ☐

Results _____

Workings: Description

Sampled by _____

Remarks

- - - - - - - - - - - - - - - ┼ - - - - - - - - - - - - - -

Sample No.   │ Sample No.

ASSAY FOR   │   ASSAY FOR

☐ Au ☐ Ag ☐ Cu ☐ Pb │ ☐ Au ☐ Ag ☐ Cu ☐ Pb
☐   │ ☐

DUPLICATE   │   ORIGINAL

GPO 878542

Figure 2

37

For the sampling of placer claims, a 16-inch gold pan is recommended and the minimum volume of material taken for a sample should be one cubic foot. The sample should be carefully panned with the heavy mineral concentrates recovered from each sample consolidated and saved separately from the concentrates of other samples. The sample concentrates will later be examined for mineral identification and the mineral content assayed.

If a pit is dug during the process of sampling, note any characteristics of the deposit such as stratification, evidences of channeling, depth below surface to water table if it is reached. Also include notes on composition of the deposit such as ratio of sand to gravel, angularity of the material, various rock types represented and those that are predominant. Check sample Form 4–1550 to be sure that all necessary data has been obtained. Include in the notes any peculiarities or pertinent data not covered by the sample form.

The formula for calculating the value per cubic yard of each drill sample of gold placers is:

$$\frac{C \times M}{A \times D} \times 27 = \text{Value per cubic yard}$$

where,

C = value of gold in cents per milligram (see figure 3)
A = area of drive shoe in square feet
D = length of sample in feet
M = milligrams of gold recovered from sample.

As most Bureau placer sampling is measured by pans of material, the above formula reduces to the following:

$$\frac{C \times M \times P}{N} = \text{value in cents per cubic yard}$$

where,

C = value of gold in cents per milligram (see figure 3)
M = milligrams of gold recovered from sample
P = number of pans per cubic yard
N = number of pans taken for sample.

A figure of 150 pans per cubic yard (P in the above formula) is generally used.

Characteristics of placer gold

Placer gold occurs as particles ranging in size from minute grains to nuggets weighing 100 or 200 pounds. Pieces worth more than 5 or 10 cents are spoken of as nuggets; smaller ones are "colors." A scale of sizes, quoted from C. F. Hoffman by Lindgren,[1] is as follows:

Coarse gold, plus 10-mesh.
Medium gold, minus 10-plus 20-mesh.
Fine gold, minus 20-plus 40-mesh.
Powder (flour) gold, minus 40-mesh.

Here, the medium gold averaged 2,200 colors per ounce, or, if

[1] Lindgren, Waldemar, *Tertiary Gravels of the Sierra Nevada of California.* (Geological Survey, U.S. Department of the Interior, Professional Paper 73, 1911.)

pure and valued at $35, about ⅔ of a color to a cent; the fine gold, 12,000 colors per ounce or 3 colors to a cent; and the powder, 400,000 colors per ounce or 10 colors to a cent. Most beach gold and some river gold, such as that of the Snake and Green Rivers, is much finer, ranging from 200 to 1,000 colors to a cent.

Colors and even nuggets almost always are flattened to some extent. Some placer gold occurs as thin flakes, which makes recovery more difficult as the flakes are not separated readily by water action from the more compact rounded grains of heavy minerals such as magnetite or garnet.

Placer gold occurs universally as an alloy with silver. Ordinarily it ranges in fineness from 700 to 950 parts of pure gold to 1,000 parts of the natural alloy, the remainder being chiefly silver. However, lower and higher degrees of fineness are common. Lindgren[2] cites the Folsom dredging field, Sacramento County, California, where the gold ranges from 974 to 978 fine. Yale[3] states that some gold from a drift mine near Vallecito, Calif., was 993 fine, or $20.52 per ounce (at $20.67), and that the gold from this property never fell below 955 fine. In a single small district the fineness of the gold is fairly uniform for any one channel. Some miners consider the fineness a distinguishing feature of a channel in districts where several channels are being explored or mined.

This rule, however, is subject to exceptions because varied sources of gold may contribute to a placer deposit and because the gold appears to lose part of its silver content and hence increases in fineness as it travels farther from its source. According to several authorities, this is due to dissolving of the silver by surface waters, an action that would have relatively more effect on fine particles than on large nuggets. Fine or flour gold is of relatively high purity.

Sand and gravel

When information on sand and gravel deposits is necessary, the sample will be sent to commercial testing laboratories. They will be instructed to conduct the various tests necessary to give the information required as to the quality of the material. Each sample should be of sufficient bulk, approximately 100 pounds, to be representative of the area involved.

---

[2] Lindgren, Waldemar, work cited, p. 68.
[3] Yale, C. G., *Gold, Silver, Copper, Lead, and Zinc in California.* (Geological Survey, U.S. Department of the Interior, Mineral Resources of the U.S., 1910, part I, p. 365.)

39

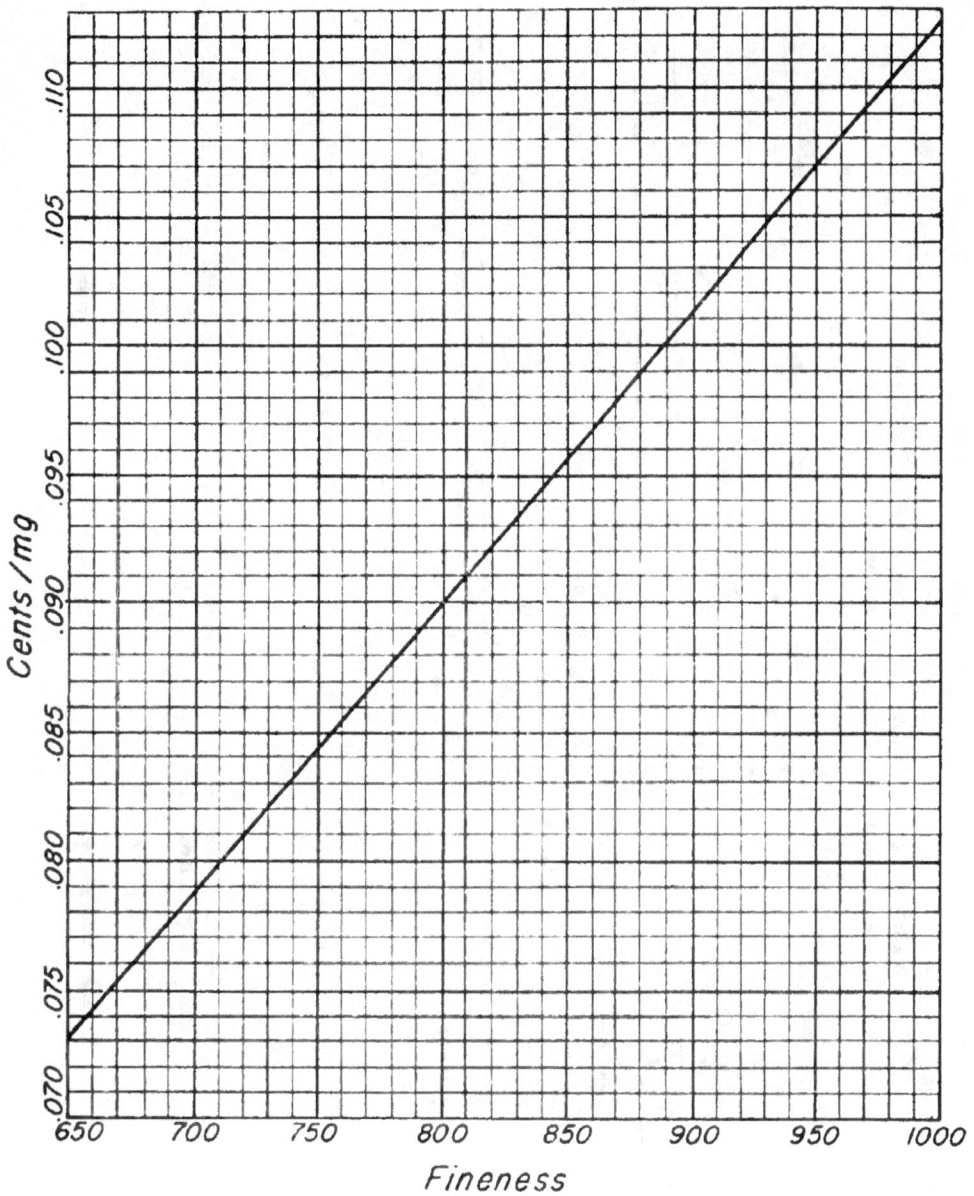

GOLD VALUE - CENTS/MG

$$M = \frac{3500 \times Fineness}{31103}$$

3500 = Cents per Ounce @ ₿ 35.00
31103 = Milligrams in Troy Oz.

Figure 3

40

Form 4—1550
(October 1960)

UNITED STATES DEPARTMENT OF THE INTERIOR
BUREAU OF LAND MANAGEMENT

PLACER SAMPLING SUMMARY

Serial Number		Sampler
Claim		Type of Sample
Sec. , T. , R. ,	M.	Date
State	County	Panner
Total depth of sample		Estimated weighted wt. of gold
Water level		Free gold assay weight
Volume panned	Cu. Yd.	Fineness of gold
Number of No. 1 colors		Cents per milligram
Average weight per color		Cents per Cu. Yd. at % recovery
Sample Notes		

Date   Engineer's Signature

Figure 4

41

DEPTH FEET	CHARACTER OF GROUND	NUMBER OF COLORS				SKETCH OF STRATA
		TRACE	1	2	3	SURFACE
	TOTAL					

No. 1 colors — 3.00 mg.   No. 2 colors — 1.33 mg.   No. 3 colors — 0.33 mg.

GPO 901072

42

Examination of County Records

The keeping of county records is not uniform from State to State or county to county within a State. There are instances where the keeping of the county records apparently depended upon the preference of the county clerk, and in some cases there were periods during which no reliable records were kept insofar as the recording of mining claims is concerned.

Occasionally county clerks may not be familiar with the manner in which the old records were kept. However, the county clerk is generally the best source of information on the keeping of the county records. Because of the lack of uniformity in the keeping of county records, it is recommended that the minerals examiner, prior to checking records, contact the county clerk or recorder and ascertain the manner in which the records in question have been kept in that county.

When the examiner is unfamiliar with the process of title searching, assistance should be requested of qualified personnel. Due to the many legal implications involved in determining present ownership of mining claims, or in obtaining a certificate of title, it is not within the scope of this discussion to attempt to lay down any rules for the performance of this work. The engineer can, however, after a careful search of available records give his opinion as to ownership. For most administrative proceedings, such as contest actions, this may be sufficient. In circumstances involving purchase of rights by negotiation or condemnation, a qualified title examiner or attorney should be consulted.

The following checklist is a guide to the records which may be available at the county court house. This list is not all inclusive nor is it intended to be a substitute for inquiry and personal familiarization of the county records by the mineral examiner.

County records

County Clerk and Recorder

1. Index books:
    a. General index: all transactions including occasionally mining deeds, grantor—grantee.
    b. Index to lodes and locators: claim name and locator; placer claims.
    c. Tract index.
    d. Abstract of mining claims.
    e. Reception or fee book.
    f. Special mining claim index books (example, Index of Affidavits of Labor).

2. General books:
   a. Mining claim locations: placer and lode.
   b. Affidavit for request of notice of publication, Public Law 585.
   c. Affidavit of labor.
   d. Transfers of mining claims (quit claim deeds).
   e. Amended location certificates.
   f. General books: contain real estate deeds.
   g. Miscellaneous books: contain all legal records (claim locations, real estate deeds, generally used in small counties).
   h. Marriage licenses.
   i. Death certificates.
   j. Mortgages.
   k. Leases.
3. County assessor:
   a. General ownership tract books.
   b. Real estate conveyance data.
   c. Copies of mineral survey plats and survey plats.
   d. Special: taxes on mining claims and improvements where claims are taxed, as well as other real estate.
4. Treasurers office.
5. County court records:
   a. Death certificates.
   b. Wills.
   c. Probate records.

Report Writing

The mineral report should be written in conformance with the format of the outline and checklist given below, keeping each chapter as concise and brief as possible while including all pertinent data. This checklist should also be used as a guide while making the field examination.

The examiner should make it a practice to use photographs and illustrations wherever possible to aid in clarifying the information contained in the report. The report should be written in the third person, keeping in mind that non-technical people should be able to read and understand it. Therefore, technical terms should be used only where necessary.

In the interest of professional ethics, the author should obtain permission to use, and should properly acknowledge the use of, data or opinions borrowed from associates. If it is necessary or desirable to use information from private sources, the author must obtain permission and, if the data is of confidential nature, the permission as well as its confidential nature must be indicated in the report. The examiner's report is not for public inspection and should be marked "For U.S. Government Use Only" (see part 442, chapter 7, Departmental Manual).

Checklist for BLM Mineral Reports

Use Standard Form 4-802 for Heading Sheet.

1. *Lands involved.*
    a. Give legal description of lands and mining claims involved.
    b. Give mineral survey number where applicable.
2. *Summary and recommendations.**
    Give citations of decisions where applicable.
3. *Introduction (when necessary).*
    a. Purpose of report.
    b. Examination made with claimant or other persons.
    c. Related pertinent factors.
4. *Status record data.*
    a. Land Office status.
    b. Location data and amendments.
    c. Property transfers.
    d. Data obtained from proofs of labor.
    e. Conflicts of land use.
    f. Prior BLM actions affecting lands such as previous decisions.
5. *Physical features.*
    a. General location of lands.
    b. How to reach the property.
    c. Description of terrain and vegetation cover.
    d. Main activity in area—mining, oil, farming, grazing, etc.
    e. Transportation, power, and water facilities.
    f. Climate.
    g. Method of identification of lands.
6. *General geology and history of area.*
    Note publications or published date.
7. *Geology and mineralization of lands involved.*
    a. Give specific details of deposition, structure size and characteristics, mineral association and occurrence.
    b. Note similarity or dissimilarity relating to favorable mineralization in surrounding areas.
8. *Mineral development work and surface improvements.*
    a. Describe in detail surface and underground workings, drill holes, including log data or other workings and exposures and their relation to the geology and mineralization.
    b. Give estimate of reasonable value and relations to mineral development of claim or lands.
    c. Describe and give present use of surplus improvements, roads, buildings, nature of occupancy.
9. *Sampling data or facts determined from marketability investigation.*
    a. Give details of sampling and relative weight attached.

---

*When necessary, insert a report index following Summary and Recommendations.

b. Give details of non-metallic prices, distances to market, haulage rates, statements or other facts ascertained for marketability investigation.
10. *Valuation (based on):*
    a. Facts.
    b. Factors relating to the economics of mining such as operating costs, labor, marketability factors.
    c. Opinions and deductions made therefrom.
    d. Factors relating to metal or mineral prices.
    e. Comparison of facts relating to use of lands for other than mineral purposes.
11. *Conclusions.*
    Summary of facts and opinions deduced therefrom, with overall conclusions based on logical analysis of problem in view of public land laws (or purpose for which report is written).
12. *Appendices.*
    a. Maps, drawings.
    b. Sketches.
    c. Assays or analysis.
    d. Photos.
    e. Statements.
    f. Historical data (old records).
    g. Title data (where necessary).
    h. Supplemental data (relating to marketability) smelter schedules, comparable sales of nonmetallics.
    i. Comparable cost information, mining, milling, and production.

Form 4–1477
(July 1959)

UNITED STATES
DEPARTMENT OF THE INTERIOR
BUREAU OF LAND MANAGEMENT

ABSTRACT OF LOCATION NOTICE

Abstract Number

Determination Area

Name of Claim

| ☐ Placer | ☐ Lode | Location Date |
| County | | Mining District |

Brief description given in notice:

Adjoining claims

LOCATORS	ADDRESSES

If above addresses were obtained from other sources, state source of information:

Cross reference to notes: Item

GPO 878508

Figure 5

47

CHAPTER 5

TESTIFYING AS AN EXPERT WITNESS

Preparation for Hearing

When the valuation engineer has completed his examination and forms a professional opinion that a mining claim is invalid because: (1) minerals have not been found within the limits of the claim(s) in sufficient quantities to constitute a valid discovery and (2) if appropriate, the land embraced within the claim(s) is nonmineral in character, a contest may be initiated.

In this type of case, an issue of fact is involved and can be resolved only as the result of a hearing. The Department has held that the Administrative Procedure Act applies to contests involving mining claims, *United States v. Keith V. O'Leary, et al.*, 63 I.D. 341 (1956) and *United States v. Keith O'Leary, et al.*, 66 I.D. 17 (1959). This act provides the means whereby both the Government and the contestee may submit both oral and written testimony at a hearing. This testimony is, of course, subject to cross-examination during the hearing by the respective attorneys. The hearing is conducted in an orderly and judicial manner by a hearing examiner at which a verbatim record of the proceeding is taken by a reporter.

When the engineer is going to testify as an expert witness, it is necessary to ascertain that he is qualified in the subject at issue. Sometimes there is need for a corroborating witness, or even an outside expert, to give emphasis or weight to support the testimony. If so, one should be used. The mining engineer can qualify as an expert witness on mining only, unless the proper foundation has been laid for him to qualify as an expert on other subjects.

The examining engineer will generally be the major witness for the Government. It is, of course, mandatory that he be well prepared to present the case at the hearing. The preparation begins by acquainting the attorney with every technical question involved in each case.

A prehearing is a "get together" of the hearing examiner with the other parties involved. Its purpose is to speed up the hearing by simplification of the issues; to exclude irrelevant issues by obtaining stipulations, admission as to facts and agreements to the introduction of documents; to limit the number of expert witnesses; and to deal with such other matters as may aid in the disposition of the proceeding.

Several days before the hearing or prehearing conference date, the engineer and solicitor, with other Government witnesses if any, should review all issues and facts of the case.

The examining engineer and the solicitor should work all of this into an orderly arrangement to be presented at the hearing. All documents, maps, pictures, assay certificates and other supporting evidence are made ready for proper submission at the hearing. If it is at all possible, the claims should be quickly looked at again to determine recent activity, if any.

In order to effectively prepare your testimony, your attention is directed to five categories of questions which counsel for an expert generally uses to develop the testimony.

1. Questions to qualify the witness as an expert.

2. Questions to associate him with the subject property and to show what he did before forming his opinion.

3. Questions to show what he considered and analyzed with respect to other data before forming his opinion.

4. Asking him to state his opinion.

5. Asking him to give reasons for the opinion he formed.

The expert should not hesitate, and where necessary should insist, on outlining his testimony in detail to his counsel before entering the courtroom. Testimony should be given in a logical sequence. In most instances this will be in chronological order, but sometimes because of other considerations it is best not to present it by point of time.

The field report is designated "For U.S. Government Use Only" and is not taken on the witness stand or introduced as evidence. Consequently, be prepared to testify from field notes and memory—know your case. However, as a memory refresher field notes or a digest of your field notes may be used. They can be used on the witness stand without introducing the notes as evidence, but they are subject to examination by the opposing counsel.

During the testimony of the contestee's witness, the engineer will listen attentively and write down any questions that should be asked by the solicitor in the cross-examination. The engineer should also assist the solicitor in the preparation of any needed rebuttal evidence.

Hearing or Court Demeanor

The most important thing to remember in giving expert testimony in court is to have a friendly, unantagonistic, unbiased attitude at all times and to try to make the judge and jury both understand that you are there, not as a witness for one side or the other, but to give your expert opinion on the value of a certain property so that an impartial and justifiable answer may be found by the jury.

An appraiser-witness when called to the stand to testify is in a sense a salesman. His task is to convey to the judge and jury his opinion of value, damages and benefits. How effectively he performs this task will depend upon the thoroughness and technique of his appraisal and the manner in which he presents his opinion in court. The following are a few of the many points of good salesmanship which, if adopted, should enhance the possibility of obtaining a verdict favorable to the opinion of the expert witness.

1. If the case is being heard by a jury, the jury should have the benefit of your answers. The answers should be given in a clear voice which can be readily heard by the jury, the judge and the court reporter. It can be quite damaging to turn your back on the jury. The witness should face the jury and direct his answer to the jury.

2. At all times have the courage and determination not to be browbeaten from what you know to be the truth or from what your convictions are. Be positive without being obstinate or stubborn.

3. In answering a question, never make a guess. If you don't know the answer, say so.

4. Do not attempt to answer questions that you do not understand. If you don't understand the question, ask the examining attorney to clarify his question.

5. Never direct your attention to your attorney as if seeking aid or assistance from him. This creates an impression of collusion between you, an impartial witness, and the attorney.

6. Information should never be volunteered in answering questions of opposing counsel. Confine your answer specifically to the question asked on cross-examination. The only exception to this is where you are absolutely certain beyond any doubt that additional information will benefit your position.

7. Don't answer the questions of opposing counsel too quickly. Never anticipate the question to be answered. Wait until it is completed before you answer. However, do not give the appearance of being hesitant. If you think the question through, your attorney will have an additional opportunity to object.

8. If you answer a question on cross-examination by a "yes" or "no" and feel that there should be a detailed explanation qualifying your answer, immediately proceed to give the explanation. If the cross-examiner attempts to stop you by saying that you have answered his question, feel free to explain to the judge why it is necessary to further enlarge on your answer.

9. Always attempt to create an impression of fairness. Answer all questions in such a manner that you do not appear to be biased. Readily admit any matters which are a fact without hesitancy even though such admission may tend to support the other side's position.

10. Keep yourself in a frame of mind so that you do not think of yourself as one of the principals in a dual. Be as polite and respectful as possible to the opposing attorney, regardless of his attitude toward you. Do not argue with the opposing attorney.

11. If questioned, frankly admit that appraising is an inexact science and that you are not dealing with exact mathematical figures.

12. Readily agree that reputable appraisers may disagree as to the value of any particular property.

13. If you make a mistake, acknowledge it. Do not attempt to squirm out of it.

14. The cross-examiner has an absolute right to examine any appraisal report or notes which the witness uses to testify on direct examination. Such examination by opposing counsel can prove embarrassing. Try not to bring to the witness stand any notes containing data which might appear contradictory, unfavorable or embarrassing. If, however, there is anything in the report or notes which you use which is adverse to your position, it should be freely brought out prior to the time it is found by the cross-examiner.

15. As a general rule, the fewer the questions asked by your attorney, the more effective your testimony will be. The lawyer may be unfamiliar with the general field and unable to phrase his questions so as to develop properly the testimony which the expert is capable of giving. If a general question is asked by your attorney, you should go into great detail and fully and explicitly answer such question.

16. Jurors are much more likely to remember anything they can see rather than something which is simply told them. The witness should use all possible illustrations, drawings, photographs, computations or other documentary evidence which can be devised to assist in the presentation of the case. The more attractive such evidence is, the more apt the jury is to remember it.

17. In presenting the income approach to value, the witness should work out, either on a blackboard or on a large sheet of blank paper, in the presence of the jury, both a round figure illustration of the capitalization method and the actual figures involved in the case on trial. If a large sheet of paper is used, it is possible that the jury may be allowed to take that illustration with them into the jury room.

18. In supporting the comparable sales approach, the appraiser should prepare a map showing the location of the comparable sales as related to the property under appraisal. Because of a divergence of views among the courts, it is advisable to identify the comparable sales on the map by number, making no reference to the names of the buyer or seller or the price paid. Before trial, provide counsel with a written list of sales which you deem comparable. This will assist in assuring proper and orderly presentation.

19. There should be nothing noted on any documentary evidence other than facts or items with which the appraiser-witness is familiar of his own knowledge.

20. You should be present and listen to the testimony of all opposing witnesses, so that you will be in a position to refute or differentiate. However, assume an air of impartiality. Do not sit at the counsel table, but take a position in the courtroom similar to that of a spectator.

21. In your direct testimony, refer to any properties which you anticipate opposing witnesses might refer to as comparable. Show in advance of such testimony why you did not consider those properties in arriving at your valuation. This should effectively steal some of the thunder of the opposition.

22. Avoid hearsay testimony as far as possible. In some courts you may be permitted to testify regarding a sale in which you did not personally participate. Generally speaking, you will be permitted (after the matter is opened up on cross-examination) to relate all the investigations upon which you based your opinion as to value, whether the information is hearsay or not. Your testimony will be more convincing if you avoid wherever possible descriptions which obviously disclose your source of information to be hearsay.

23. Remember that you are testifying under oath. Don't say or do anything that would lead the jury to believe that you are falsifying, misleading or concealing anything.

24. Refer to other witnesses respectfully and in a friendly and considerate manner. Try to convince the jury that you know what you are doing, that you are fair and sincere, and that you are not infallible.

25. If a juror is permitted to ask you a question, regardless of how irrelative it may seem, take great care to give a thorough and fair answer and explanation. If the judge asks the question, answer in a way that recognizes his position and the importance of what he asks. Make certain that he feels that you have answered him thoroughly.

26. Never appear hurried or bored. Act as though this case is the most important business you have, deserving of careful, considered treatment.

27. Avoid technical terms. Speak in the language of the layman juror.

28. In all your testimony, lean over backwards to give the landowner the benefit of the doubt. Round your figures in favor of the landowner.

29. If anyone is to get irritated, bored, impatient, angered, disgusted, let it be your lawyer. Never you.

30. Above all, do be prepared.

Left chart:

Era	Period (or system)	Epoch (or series)		
Cenozoic	Quaternary	Recent		
		Pleistocene ("Glacial")		
	Tertiary	Pliocene		
		Miocene		
		Oligocene		
		Eocene		
		?		
Mesozoic	Cretaceous	Upper Cretaceous	Gulf series (Provincial series in southwestern US)	
		Lower Cretaceous	Comanche series (Provincial series in southwestern U.S)	Shasta series (Provincial series of Pacific coast)
	Jurassic	Upper Jurassic		
		Middle Jurassic		
		Lower Jurassic		
	Triassic	Upper Triassic		
		Middle Triassic		
		Lower Triassic		

Right chart:

Era	Period	Epoch	
Paleozoic	Carboniferous	Permian (Contains workable beds of coal; fauna closely related to Pennsylvanian fauna)	
		Pennsylvanian (The great coal-bearing series commonly known as "Coal Measures")	
		Mississippian (Contains workable beds of coal in some areas)	
	Devonian	Upper Devonian	
		Middle Devonian	
		Lower Devonian	
	Silurian		
	Ordovician	Upper Ordovician (Cincinnatian, provincially)	
		Middle Ordovician (Mohawkian, provincially)	
		Lower Ordovician	
	Cambrian	St. Croixan or Upper Cambrian	
		Acadian or Middle Cambrian	
		Waucoban or Lower Cambrian	
Proterozoic	Algonkian (Few fossils)	Keweenawan	Belt: Mont. Idaho. Grand Canyon: Ariz. Glenarm: S.E.Pa. Md., Va. Grenville: N.Y. Llano: Tex. (Provincial series; not chronologic succession)
		Huronian	
Archean		Laurentian (Intrusive)	
		Keewatin (No undisputed fossils)	

Figure 6

53

G. Schuchert (Text-Book Geol., 2d ed., pt. 2, 1924)			
Era	Major chron.	Period	Epoch
Psv. cha. chu.		Recent or post-Glacial	
Cenozoic	Neogene		Pleistocene or Glacial
			Pliocene
			Miocene
	Paleogene		Oligocene
			Eocene
Mesozoic	Late Mesozoic	Upper Cretaceous	Fort Union
			Lance
			Montanian
			Coloradian
		Lower Cretaceous	(Cenomanian at top)
	Early Mesozoic	Jurassic	Upper Jurassic or Malm
			(Callovian at base)
			Middle Jurassic or Dogger
			Lower Jurassic or Lias
		Triassic	Upper Triassic
			Middle Triassic
			Lower Triassic

Paleozoic	Late Paleozoic or Carboniferous	Permian		
		Pennsyl-vanian		
	Middle Paleozoic	Missis-sippian	Tennesseian	
			Waverian (Chattanooga, base)	
		Devonian	Upper Devonian	
			Middle Devonian	
			Lower Devonian (Manlius at base)	
	Early Paleozoic	Silurian	Cayugan	
			Niagaran	
			Alexandrian	
		Ordovician	Upper Champlainian	Gamache
				Richmond
				Maysville-Lorraine
				Eden
			Middle or Mohawkian	
			Lower Champlainian	Chazyan
				Beekmantown
		Cambrian	Ozarkian	
			Croixian	
			Acadian	
			Taconian	
Proterozoic (Late)	Algonkian		Killarney revolution (granite)	
			Keweenawan	Beltian
			Animikian	
			Huronian	
Proterozoic (Early)	Timiskamian		Algoman revolution (granites)	
			Sudburian-Doréan d	
Archeozoic or Archean			Laurentian revolution	
	Loganian		Keewatin-Coutchiching	Grenville e

54

GENERALIZED DIAGRAMS OF THE
RECTANGULAR SYSTEM OF SURVEYS

TOWNSHIP GRID

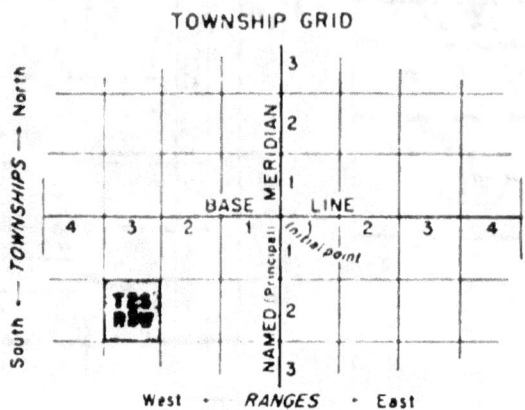

TOWNSHIP 2 SOUTH, RANGE 3 WEST

6	5	4	3	2	1
7	8	9	10	11	12
18	17	16	15	14	13
19	20	21	22	23	24
30	29	28	27	26	25
31	32	33	34	35	36

SECTION 14

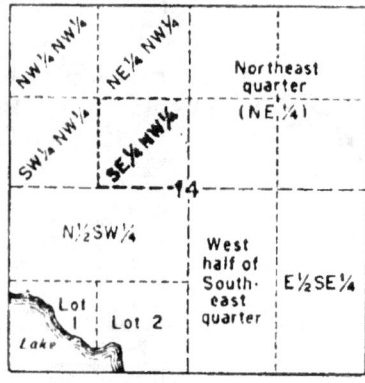

= Location and Description of Land Tracts =

Under this survey plan any tract can be located and described from its coordinate position within its principal meridian. For example, if the shaded areas were controlled by the Boise Meridian, the abbreviated description of the smallest subdivision would be the

av   SE¼ NW¼, Sec 14, T 2 S, R 3 W, Boise Mer

Figure 7

55

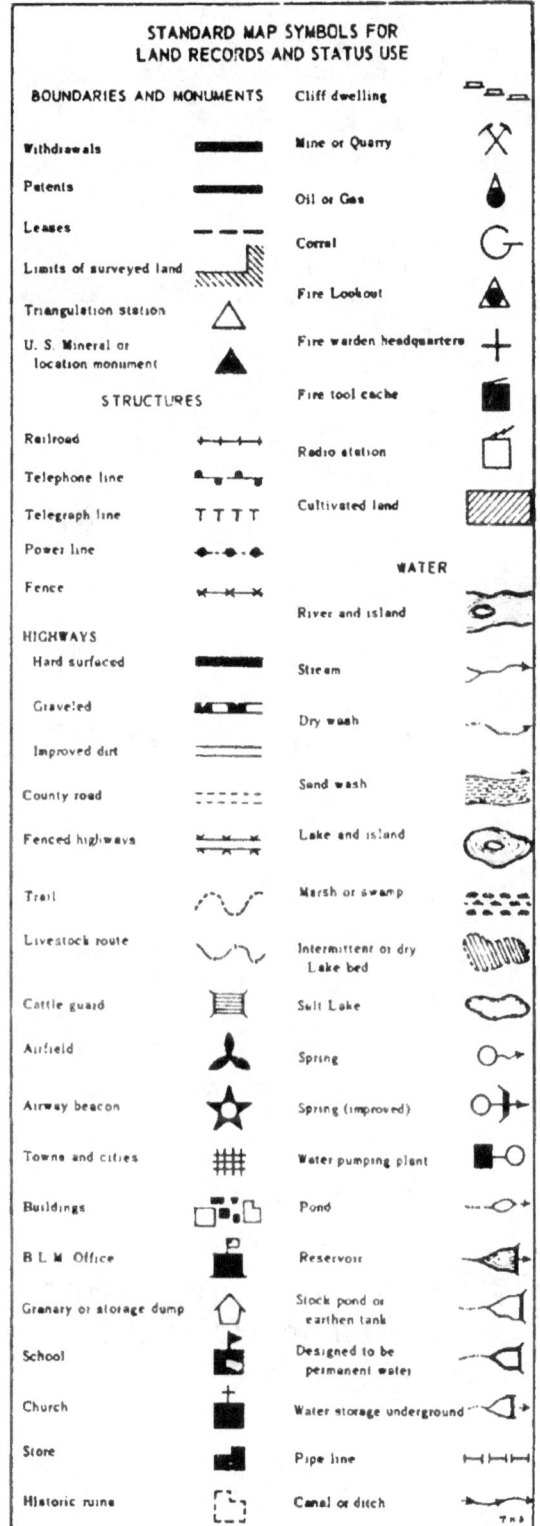

Figure 8

56

MARKINGS FOR STONE
SECTION CORNERS

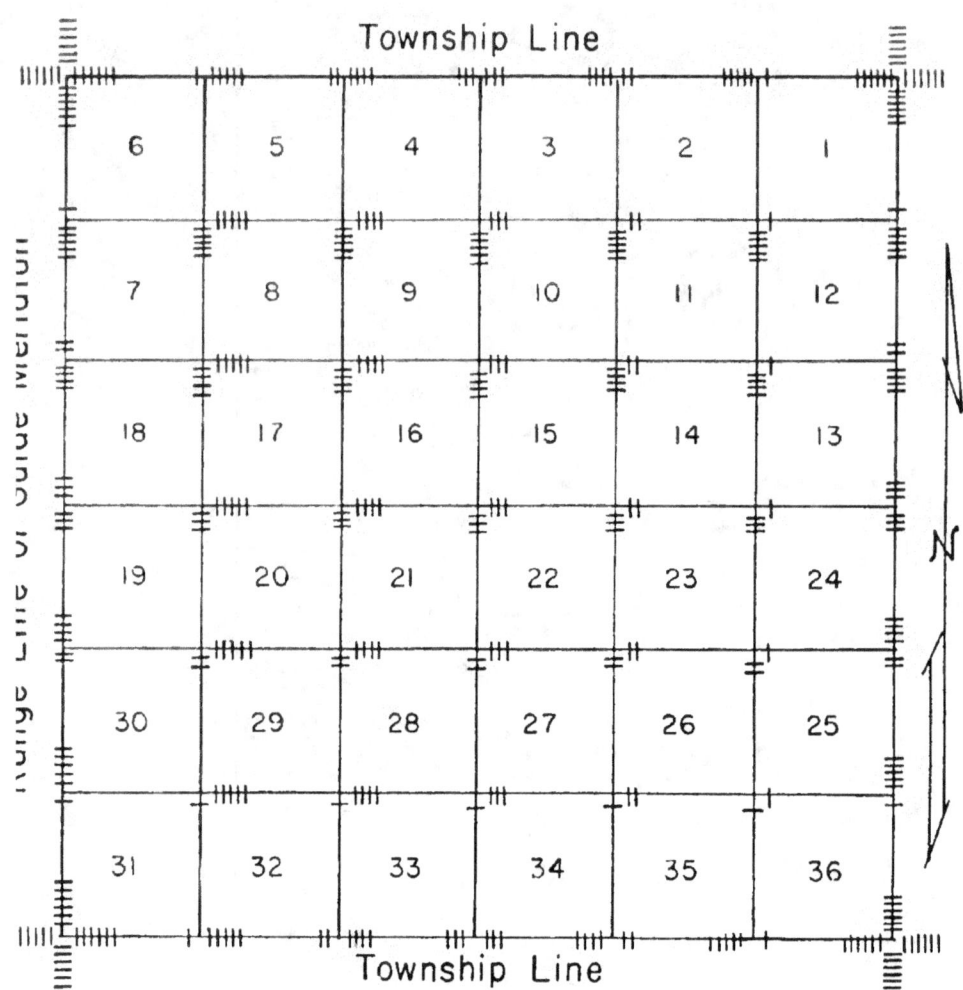

Quarter-Section Corners are marked with the fraction "$\frac{1}{4}$", those on meridional lines on their west, and those on latitudinal lines on their north faces.

Figure 9

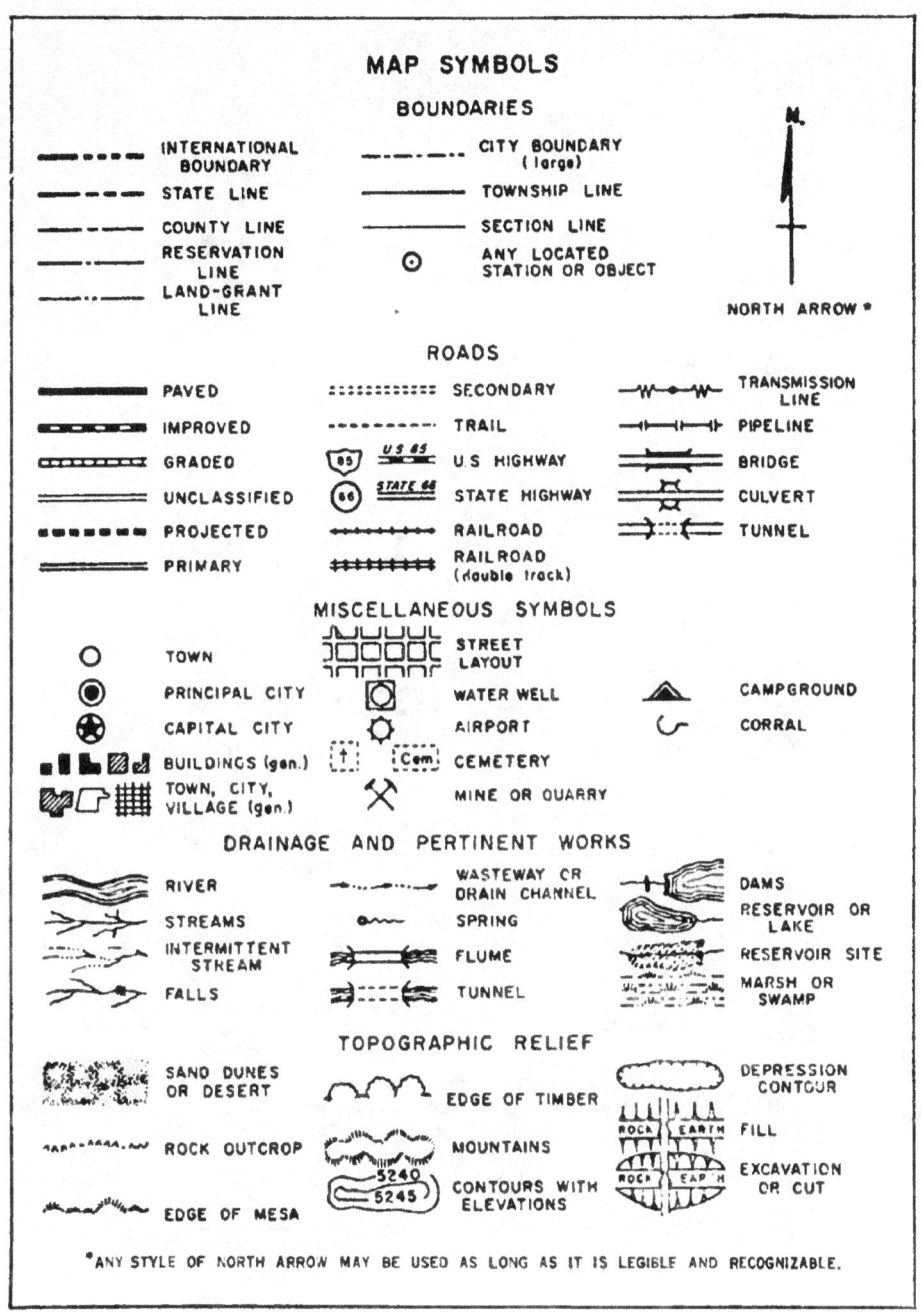

Standard Map Symbols.

Figure 10

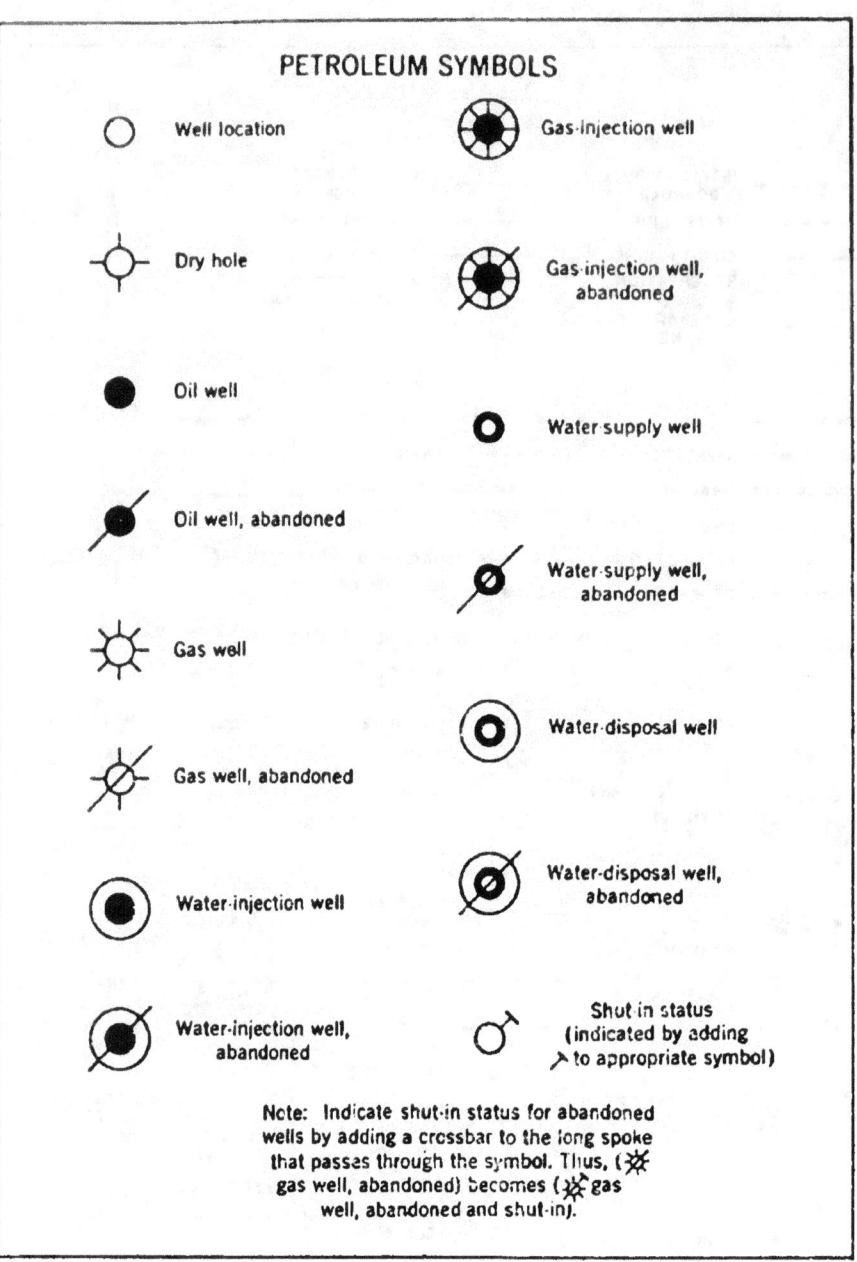

Standard Petroleum Symbols.

Figure 11

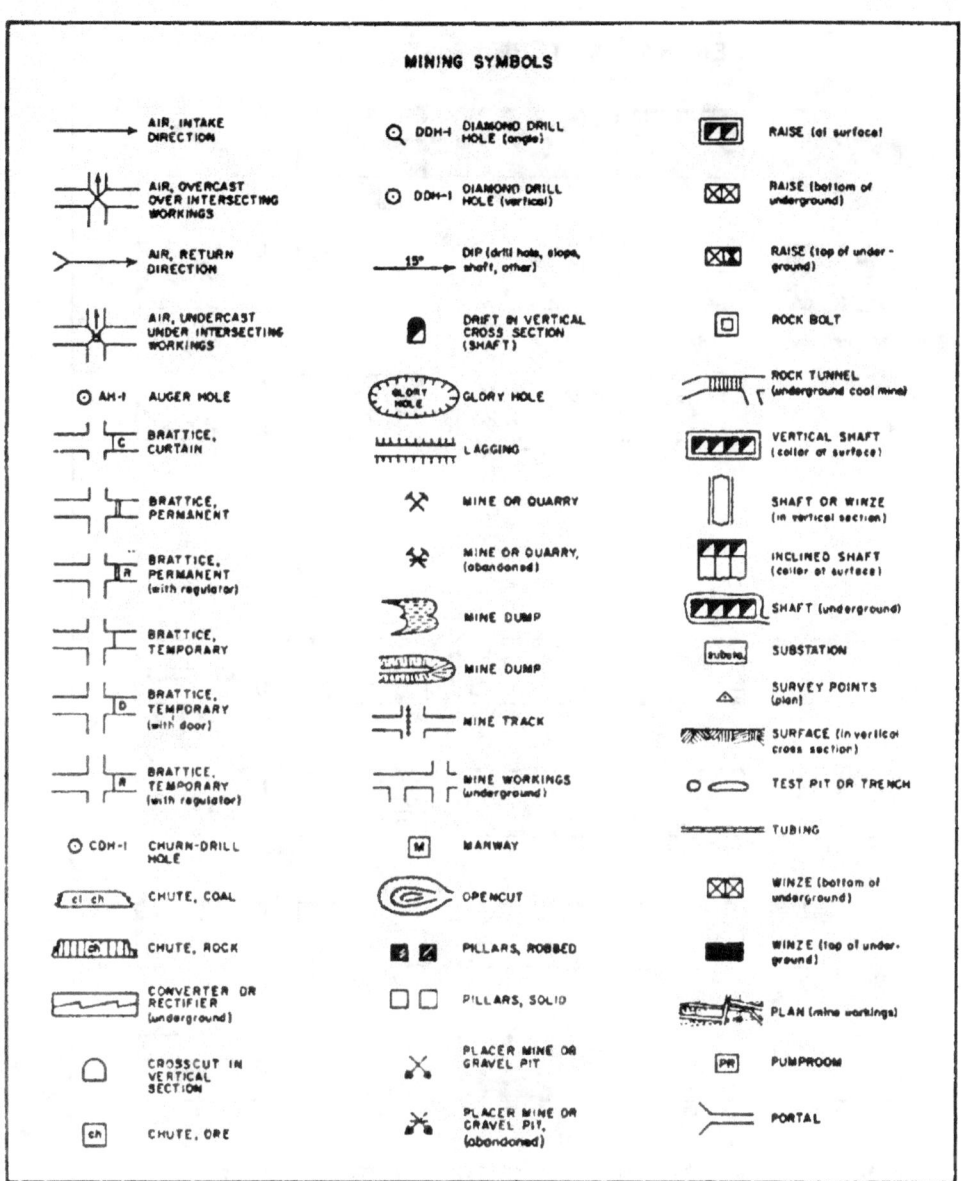

Standard Mining Symbols.

Figure 12

60

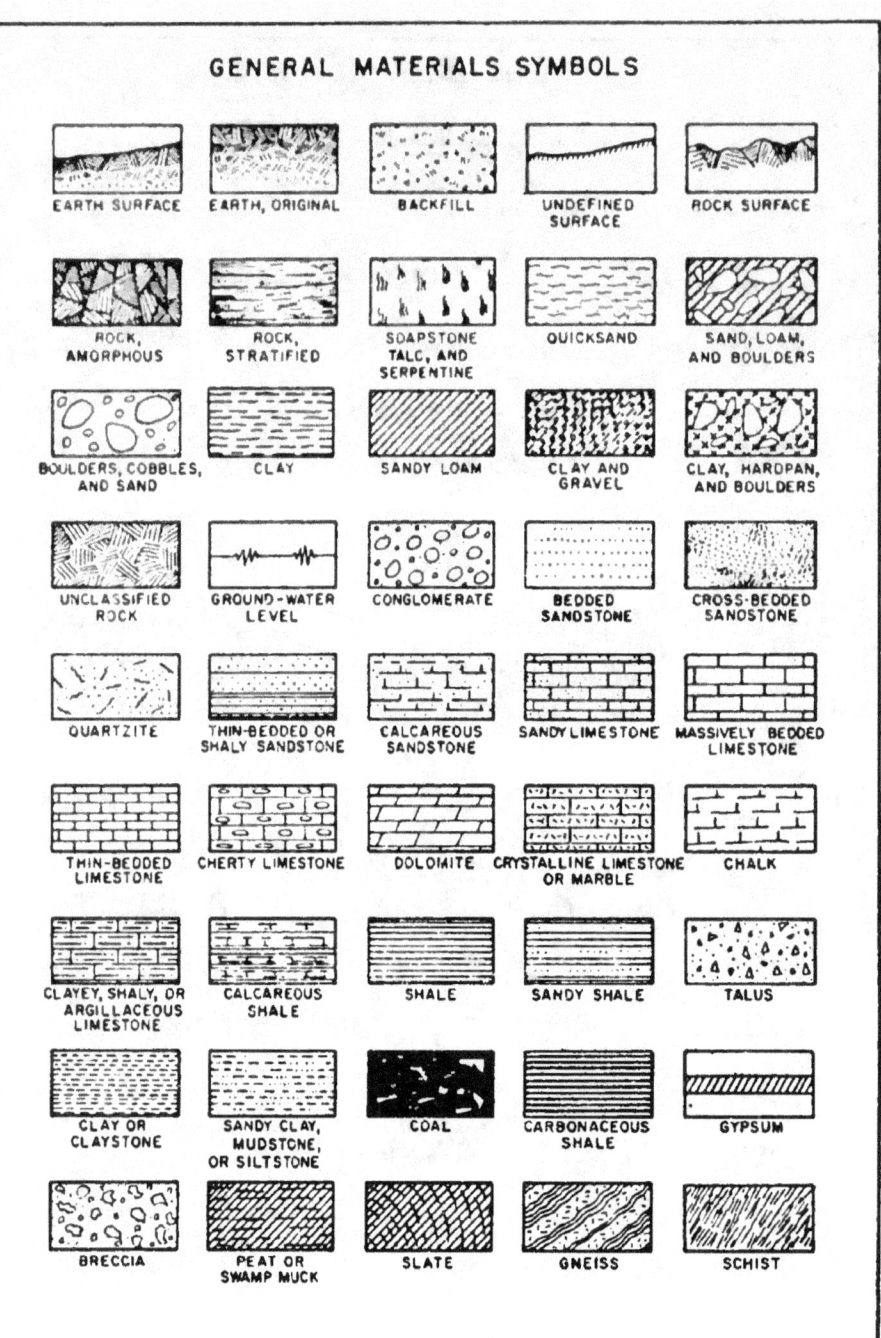

Standard Materials Symbols.

Figure 13

61

MAP SYMBOLS FOR GEOLOGIC
AND MATERIALS INVESTIGATIONS

DESCRIPTION	PROPOSED	COMPLETED
Drill hole (up to and incl. 6")	⊕ DH-2	⬤ DH-2
Drill hole, large dia. (above 6")	⊕ DH-2	⬤ DH-2
Angle drill hole	⊕ DH-2	⬤ DH-2
Auger hole (up to and incl. 6")	○ AH-2	● AH-2
Auger hole, large dia. (above 6")	◎ AH-2	◉ AH-2
Test pit	▢ TP-2	◨ TP-2
Test shaft	◺ TS-2	◢ TS-2
Test trench	⟋ TT-2	▬ TT-2
Test drift	⟨ TD-2	◣ TD-2

When advantageous, indicate special character of drill hole or sampling
by these abbreviations in parentheses, following drill hole number.

DESIGNATIONS	EXAMPLES
DN = Denison samples	◉ DH-2(DN)
DS = Drive-sample hole	⊕ DH-3(DS)
CX = Calyx-drill hole	◉ DH-4(CX)
CD = Churn-drill hole	⊕ DH-5(CD)
WB = Wash boring	⬤ DH-6(WB)

Standard Symbols for Geologic and Materials Investigations.

GPO 840243

Figure 14

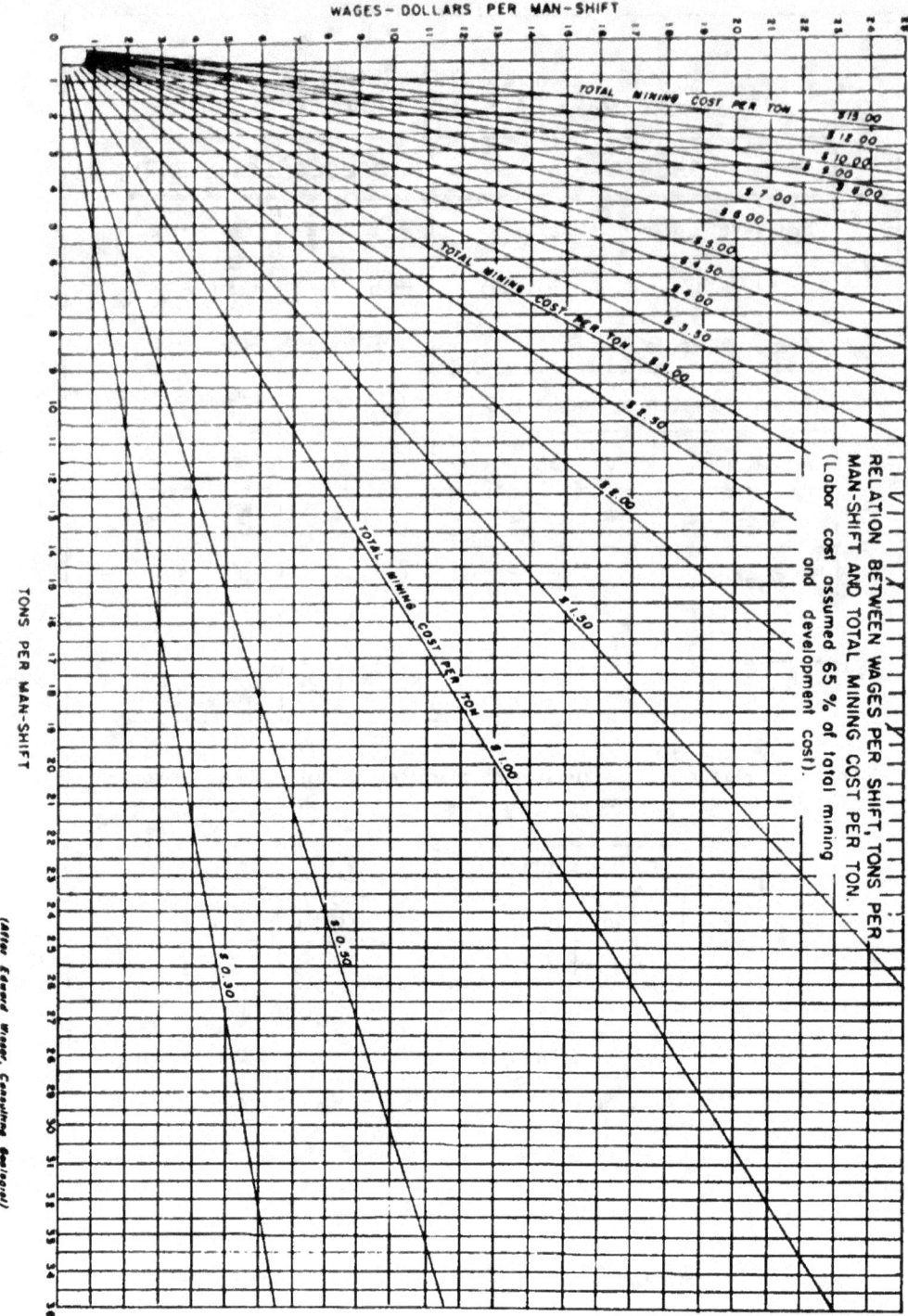

Figure 15

63

Acknowledgments

Tables 1, 12, 18, 19, and 20 have been reprinted with permission of Denver Equipment Company, Denver, Colo.
Tables 2, 3, 4, 5, and 6 reprinted from *Economic Mineral Deposits*, by A. M. Bateman (John Wiley and Sons, Inc., 1950).

TABLE 1.—*Minerals and*

Name	Formula	Percent metal	Color	Lustre
Actinolite	$Ca(MgFe)_3 (SiO_3)_4$.	No metal source.	Green	Vitreous
Albite	$NaAlSi_3O_8$	Al_2O_3—19.5%	White to blue	Vitreous
Almandite	$Fe_3Al_2(SiO_4)_3$	No metal source.	Red to black	
Altaite	$PbTe$	61.9% Pb	Tin white, yellow tinge.	Metallic
Alunite	$K_2(Al_2OH)_6\cdot (SO_4)_4$.	K—9.4% Al—19.6%.	Pink-red	Vitreous pearly.
Amosite	$(FeCaH_2Mn) OSiO_2$.	No metal source.	Gray to green	
Analcite	$NaAlSi_2O_6\cdot 2H_2O$.	Al_2O_3—23.2%	White	Vitreous
Andalusite	Al_2SiO_5	Al_2O_3—63.2%	White, red-green.	Vitreous
Andradite	$Ca_3Fe_2(SiO_4)_3$	No metal source.	Green, red-black.	Adamantine
Anglesite	$PbSO_4$	Pb—68.3%	Yellow, green-gray.	Adamantine, vitreous.
Anorthite	$CaAl_2Si_2O_8$	Al_2O_3—36.7%	White, gray-red.	Vitreous
Anthophyllite	$(MgFe)SiO_3$	No metal source.	Gray, brown-green.	Vitreous
Apatite	$Ca_4(CaF) (PO_4)_3$.	P_2O_5—42.3%	Green-blue	Vitreous
Aragonite	$CaCO_3$	CaO—56%	White	Vitreous
Argentite	Ag_2S	Ag—87.1%	Black	Metallic
Argyrodite	$3Ag_2S\cdot GeS_2$	Ag—73.5%	Steel gray, red tinge.	Metallic
Arsenopyrite	$FeAsS$	Fe—34.3% As—46.0%	Steel gray	Metallic
Atacamite	$Cu_2(OH)_3Cl$	Cu—59.5%	Green	Adamantine, vitreous.
Azurite	$2CuCO_3\cdot Cu(OH)$.	Cu—55.0%	Blue	Vitreous, dull
Barite	$BaSO_4$	BaO—65.7%	White, blue-red.	Vitreous
Bauxite	$Al_2O_3\cdot 3H_2O$	Al—34.9%	White-red, brown-yellow.	Dull
Bentonite	$(CaMg) O\cdot SiO_2 (AlFe)_2O_3$.	No metal source.	Blue	Dull

Acknowledgments—Continued

Table 9 from *Mining Geology*, by Hugh E. McKinstry (Prentice-Hall, Inc., Englewood Cliffs, N.J., 1948).

Table 16 reprinted with permission from *Mining Engineer's Handbook*, by R. Peele (John Wiley and Sons, Inc., 1945).

their characteristics

Streak	Hardness	Specific gravity	Characteristics—occurrence
-----------	5.0–6.0	3.0–3.2	Usually long crystals, columnar or fibrous.
White-------	6.0–6.6	2.6–2.7	Occurs sometimes in platy masses. Otherwise like anorthite. See anorthite.
White-------	6.5–7.5	3.1–4.3	Variety of garnet. See garnet.
Grayish black---	3.0	8.2	Associated with pyrite, galena, tetrahedrite.
White-------	3.8	2.7	Associated with kaolin and pyrite.
-----------	-----------	2.2–2.3	Long fibered asbestos.
White-------	5.0–5.5	2.2–2.3	Trapezohedral crystals in cavities in basic igneous rocks.
-----------	7.5	3.2	Nearly square prisms; occurs with gneiss, mica, schists.
White-------	6.5–7.5	3.1–4.3	Variety of garnet. See garnet.
White-------	2.8–3.0	6.1–6.4	Occurs in oxidation zones of lead veins.
White-------	6.0–6.5	2.7–2.8	Tabular crystals in igneous rocks, with fine longitudinal lines on the better of two perfect cleavages at 90° to each other.
Uncolored, grayish.	5.0	3.0–3.2	Found in crystalline schists.
White-------	4.5–5.0	3.2	Usually granular or in 6-sided prisms.
White-------	3.5–4.0	2.9	Effervesces like calcite. Powder becomes lilac or purple when boiled in 10X solution of cobalt nitrate.
Shiny black-----	2.0–2.5	7.2–7.4	Cuts like lead; with silver, cobalt and nickel.
Grayish black---	2.5	6.1	Occurs with sphalerite, siderite and marcasite.
Gray, black-----	5.5–6.0	5.9–6.3	Widely spread; yields sparks and garlic odor when struck slanting blows with steel.
Apple green-----	3.0–3.5	3.8	Always of secondary origin with copper ores.
Blue-----------	3.5–4.0	3.8–3.9	Oxidized mineral that effervesces vigorously in muriatic acid of any strength and temperature.
White-------	2.5–3.5	4.3–4.6	Found commonly as gangue of lead-zinc ores. Platy or granular masses or either diamond-shaped or rectangular crystals.
Like color-------	1.0–3.0	2.6	Chief ore of aluminum; occurs massive. Completely soluble in salt of phosphorous bead.
Light gray------	1.0	2.1	The clay of montmorillonite. Swells greatly when placed in water.

TABLE 1.—*Minerals and their*

Name	Formula	Percent metal	Color	Lustre
Beryl	Be₃Al₂(SiO₃)₆	Be—5% Al₂O₃—19%.	White, green-blue.	Vitreous
Beryllonite	NaBePO₄	Be—7.1%	White-yellow	Vitreous, brilliant.
Biotite	(HK)₂ (MgFe)₂ Al₂(SiO₄)₃.	No metal source.	Black-Brown	Pearly, vitreous.
Bismite	Bi₂O₃	No metal source.	Straw yellow, white.	Pearly
Bismuth	Bi	Bi—100%	Silver White	Metallic
Bismuthinite	Bi₂S₃	Bi—81.2%	Lead gray	Metallic
Bismutite	(BiO)₂·CO₃·H₂O.	No metal source.	Green-white	Vitreous, dull
Borax	Na₂B₄O₇·10H₂O.	B₂O₃—36.6% Na₂O—16.2%.	White	Vitreous, dull
Bornite	Cu₃FeS₄	Cu—63.3%	Reddish	Metallic
Bournonite	3(PbCu₂) S.Sb₂S₃.	Pb—24.7% Cu—42.5%.	Steel gray, iron black.	Metallic
Braunite	3Mn₂O₃·MnSiO₃.	Mn—78.3%	Steel gray, brownish black.	Submetallic
Breithauptite	NiSb	Ni—32.5% Sb—67.5%.	Copper red	Metallic
Brochantite	CuSO₄·3Cu(OH)₂	Cu—56.2%	Green	Vitreous
Brucite	MgO·H₂O	MgO—69%	White to gray, blue, green.	Pearly, vitreous.
Calamine	H₂(Zn₂O)·SiO₄.	ZnO—67.5%	White, blue, green, brown.	Vitreous, dull
Calaverite	AuTe₂	Au—43.6%	Bronze yellow, silver-yellow tinge.	Metallic
Calcite	CaCO₃	CaO—56%	Many colors	Vitreous
Calomel	HgCl	Hg—85% Cl—15%.	White, yellow	Adamantine
Carnallite	KMgCl₃·6H₂O.	K—14.1% Cl—38.3%.	White	Shining
Carnotite	K₂O·2U₂O₃·V₂O₅·3H₂O variable.	Variable	Yellow	Vitreous, dull
Cassiterite	SnO₂	Sn—78.8%	Brown, black, red.	Adamantine
Celestite	SrSO₄	Sr—47.7%	Light blue, white, red.	Vitreous
Cerargyrite	AgCl	Ag—75.3%	Pearly gray	Waxy, greasy
Cerussite	PbCO₃	Pb—77.5%	White, gray	Adamantine
Cervantite	2Sb₂O₄ Sb₂O₃·Sb₂O₅.	Sb—79.4%	Yellow reddish white.	Greasy, pearly.
Chalcanthite	CuSO₄·5H₂O	CuO—31.8%	Blue	Vitreous

66

Streak	Hardness	Specific gravity	Characteristics—occurrence
White	7.5-8.0	2.6-2.8	Often imbedded in quartz; with mica, feldspar. Usually in 6-sided prisms with flat terminations in pergmatite. Gem varieties.
White	5.8	2.8	Found with beryl, feldspar, columbite.
White	2.5-3.0	2.7-3.1	Cleaves easily into very thin, flexible and elastic plates.
Like color		4.4	Of secondary origin resulting from oxidation.
Silver white	2.3	9.7	Native; with cobalt, nickel; brassy tarnish.
Like color	2.0	6.4-6.5	Occurs in form of thin coating.
Greenish gray-white.	4.0	6.9-7.7	Incrusting fibrous, or earthy and pulverulent.
White	2.0-2.5	1.7	Refer to introduction for characteristic taste.
Blackish gray	3.0-3.5	4.9-5.4	Associated with chalcoite. Usually massive. Quickly tarnishes iridescent blue.
Like color	2.5-3.0	5.7-5.9	Occurs fine-grained massive; brittle.
Like color	6.0-6.5	4.8	Occurs in porphyry; brittle.
Reddish brown	5.5	7.5	Occurs with other sulfides and silver minerals.
Green	3.5-4.0	3.9	Oxidized mineral. Dissolves quietly in nitric acid.
White	2.5	2.4	Associated with serpentine; secondary mineral.
White	4.5-5.0	3.4-3.5	Usually in crystal coatings; sometimes in cockscomb-like aggregates. Often with smithsonite.
Yellowish gray	2.5	9.0	Similar to sylvanite, but never in crystals.
White	3.0	2.7	Massive and 6-sided pointed or prismatic crystals. Effervesces vigorously in muriatic acid of any strength or temperature.
Pale yellow, white.	1.0-2.0	6.5	Associated with cinnabar.
White	2.5	1.6	Strongly phosphorescent; taste—bitter.
Yellow	1.5		Powder or earth in sandstone. Often concentrated around petrified wood.
White, light brown.	6.0-7.0	6.8-7.1	Massive or squarish crystals.
White	3.0-3.5	3.9-4.0	Same as barite.
White to gray	1.0-1.5	5.6	Cuts like wax; exposure changes color to violet brown.
White	3.0-3.5	6.5-6.6	Oxidized mineral. Effervesces vigorously in warm concentrated or boiling dilute muriatic acid.
White	4.0-5.0	4.1-5.3	Usually associated with stibnite.
White	2.5	2.1-2.3	Oxidized mineral. Tastes metallic and nauseating.

TABLE 1.—*Minerals and their*

Name	Formula	Percent metal	Color	Lustre
Chalcedony	SiO_2	No metal source.	Pale blue, gray, white to black.	Waxy
Chalcocite	Cu_2S	Cu—79.8%	Black-gray	Metallic
Chalcomenite	$CuSeO_3 \cdot 2H_2O$	Cu—28.1% Se—34.9%.	Blue	Vitreous
Chalcopyrite	$CuFeS_2$	Cu—34.6%	Brassy yellow	Metallic
Chert	SiO_2	No metal source.	White-gray	Dull
Chloanthite	$NiAs_2$ variable.	Ni—28.1% As—71.9%.	Tin white, steel gray.	Metallic
Chromite	$FeO \cdot Cr_2O_3$	Cr—46.2%	Black	Vitreous
Chrysoberyl	$BeOAl_2O_3$	BeO—19.8%	Green	Vitreous
Chrysocolla	$CuOSiO_2 \cdot 2H_2O$.	Cu—36.2%	Blue, green	Vitreous, dull
Chrysolite	$(MgFe)_2SiO_4$	No metal source.	Green	Vitreous
Chrysotile	$H_4Mg_3Si_2O_9$		White, greenish.	Metallic
Cinnabar	HgS	Hg—86.2%	Red	Adamantine, submetallic.
Clausthalite	$PbSe$	Pb—72.4%	Lead gray	Metallic
Cobaltite	$CoAsS$	Co—35.5%	Tin white, steel gray.	Metallic
Colemanite	$Ca_2B_6O_{11} \cdot 5H_2O$.	No metal source.	White, yellowish.	Brilliant, vitreous.
Columbite	$(FeMn)(CbTa)_2O$.	Variable— Ta_2O_5 3.3 to 31.5%.	Iron black	Submetallic
Copper	Cu	Cu—100%	Copper red	Metallic
Corundum	Al_2O_3	Al—52.9%	All colors	Vitreous, adamantine.
Cosalite	$Pb_2Bi_2S_5$	Pb—41.8% Bi—42.1%.	Lead gray	Metallic
Covellite	CuS	Cu—66.5%	Blue	Submetallic
Crocidolite	$NaFe(SiO_3)_2 \cdot FeSiO_3$.	No metal source.	Blue to green	Silky, dull
Crocoite	$PbCrO_4$	Pb—64.1% Cr—16.1%.	Red	Adamantine
Cryolite	Na_3AlF_6	Al—13% F—54.4%.	Snow white	Greasy to vitreous.
Cuprite	Cu_2O	Cu—88.8%	Red	Adamantine to dull.
Cyanite (or kyanite).	Al_2SiO_5	Al—33.3%	White, to blue or green.	Vitreous, pearly.
Desclolzite	$4RO.V_2O_5.H_2O$	Variable— V_2O_5.	Red, brown, black.	Greasy
Diamond	C	C—100%	White, gray	Adamantine, greasy.
Diaspore	$Al_2O_3 \cdot H_2O$	Al_2O_3—85%	Many colors	Vitreous
Diatomaceous earth.	$SiO_2 \cdot nH_2O$		Yellow to brown.	Vitreous
Dolomite	$CaMg(CO_3)_2$	CaO—30.4% MgO—21.9%.	White, gray, pink, yellow.	Vitreous, pearly.

Streak	Hardness	Specific gravity	Characteristics—occurrence
White	7.0	2.6–2.7	Smoothly rounded fracture. Semi-precious gem varieties.
Like color........	2.5–3.0	5.5–5.8	Highly polished surface where cut.
Bluish white....	2.5–3.0	3.8	With various selenides of silver, copper, and lead.
Greenish black..	3.5–4.0	4.1–4.3	Softer than pyrite; with pyrite, galena, sphalerite.
White..........	7.0	2.6	Impure, coarse-grained, opaque flint.
Grayish black...	5.8	6.5	Granular or in crystals like pyrite. Often associated with erythrite. See erythrite.
Dark brown....	5.5	4.3–4.6	Grains may look like black glass. Often with serpentine.
White..........	8.5	3.7–3.8	Usually in crystals or worn pebbles. Gem varieties.
White..........	2.0–4.0	2.0–2.2	Adheres to dry tongue; important ore of copper.
White or yellowish.	6.5–7.0	3.3	In granular masses, glassy grains or crystals. Gem varieties.
White..........	1.7	2.2	Best asbestos. Masses of tough, usually parallel, slender fibers.
Scarlet..........	2.0–2.5	8.0–8.2	Only important ore of mercury; tastes chalky.
Lead gray.......	2.8	8.0	Resembles granular galena.
Grayish black...	5.5	6.0–6.3	Granular or in crystals like pyrite. Often with erythrite. See erythrite
White..........	4.0–4.5	2.4	Usually occurs as geodes; brittle.
Dark red, black.	6.0	6.3	Brittle; nearly pure niobate.
Copper-red.....	2.8	8.8	Tarnishes easily; malleable.
White..........	9.0	3.9–4.1	In 6-sided crystals and masses that may break in three directions at nearly 90°. Gem varieties.
Black..........	2.8	6.5	In quartz veins; with pyrite, sphalerite.
Black..........	1.5–2.0	4.6	Platy or granular massive. Turns purple when moistened.
Like color.......	4.0–5.0	3.2–3.3	Fibrous masses; like asbestos, valuable.
Orange yellow..	2.5	6.0	Found with quartz, galena, vanadinite.
White..........	2.5	3.0	Appearance, hardness are distinctive.
Red............	3.5–4.0	5.9–6.2	Oxidized mineral. Often in crystals— usually cubical.
....	4.0–7.0	3.6	Bladed crystals with flat cleavage surfaces that are easily scratched longitudinally but not transversely.
Orange..........	3.5	6.0	Associated with vanadinite.
Ash gray........	10.0	3.5	Occurs in crystals (usually rounded octahedrons) in a basic igneous rock, and in placers. Gem varieties.
White..........	6.5–7.0	3.4	Occurs in thin scales; very brittle.
White to gray...	2.0	2.2	Roughens glass. Uniformly very fine texture and light in weight.
White..........	3.5–4.0	2.8–2.9	Effervesces vigorously in any condition of muriatic acid except cold dilute. Like calcite, but common in warped rhombohedrons.

TABLE 1.—*Minerals and their*

Name	Formula	Percent metal	Color	Lustre
Enargite	$3Cu_2S·A_2S_5$	Cu—48.4%	Iron black	Metallic
Epidote	$Ca_2(AlOH)$ $(AlFe)_2$ $(SiO_4)_3$.	No metal source.	Green	Vitreous, dull
Epsom salt	$MgSO_4·7H_2O$	Mg—9.9%	White	Vitreous
Erythrite	$Co_3As_2O_8$ $·8H_2O$.	Co—29.5%	Usually pink, gray.	Pearly
Ferberite	$FeWO_4$	W—60.6%	Brown, black, gray.	Metallic
Fluorite	CaF_2	F—48.9%	All colors	Vitreous
Franklinite	$(ZnFeMn)O$ $(FeMn)_2O_3$.	Zn—14.2% Mn—35.7%	Iron black	Metallic
Galena	PbS	Pb—86.6%	Lead gray	Metallic
Garnet	Various	No metal source.	Red, brown yellow.	Vitreous
Garnierite	$H_2(NiMg)$ SiO_4.	Ni—25% to 30%.	Green	Dull, greasy
Genthite	$2NiO,2MgO,$ $3SiO_2·6H_2O$.	Ni—22.6%	Green	Dull, greasy
Gibbsite	$Al(OH)_3$	Al—34.6%	White, green	Pearly
Gold	Au	Au—100%	Golden	Metallic
Graphite	C	C—100%	Black	Dull, submetallic.
Greenockite	CdS	Cd—77.7%	Yellow	Adamantine
Grossularite	$Ca_3Al_2(SiO_4)_3$	No metal source.	White, green, yellow.	Vitreous
Gypsum	$CaSO_4·2H_2O$	CaO—32.6%	White, red	Vitreous
Halite	NaCl	Na—39.4%	White	Vitreous
Halloysite	$H_4Al_2O_3·2SiO_2$ $·H_2O$.	No metal source.	White, green, blue, red.	Pearly, waxy, dull.
Hausmannite	Mn_3O_4	Mn—72%	Black, brown	Metallic
Hematite	Fe_2O_3	Fe—70%	Brown, red, black.	Metallic, dull, submetallic.
Hessite	Ag_2Te	Ag—63%	Gray	Metallic
Horneblende	Variable	Variable	White, green, black.	Vitreous
Huebnerite	$MnWO_4$	Mn—18.1% W—60.7%	Brown	Submetallic
Hydrozincite	$ZnCo_3$ $·2Zn(OH)_2$.	Zn—59.5%	White, gray, yellow.	Dull
Hypersthene	$(FeMg)SiO_3$	No metal source.	Black	Pearly
Ilmenite	$FeT O_3$	Ti—C1.6%	Iron black	Metallic, submetallic.
Iodyrite	AgI	Ag—46%	Yellow, green	
Iridium	Variable	Alloy—100%	White	Metallic
Iridosmene	IrOs $(RhPtRu)$.	Alloy—100%	Tin white	Metallic
Jamesonite	$2PbS·Sb_2S_3$	Pb—50.8% Sb—29.5%	Gray	Metallic

70

Streak	Hardness	Specific gravity	Characteristics—occurrence
Black_____	3.0	4.4	Color and streak both black; prismatic cleavage.
White_____	6.0-7.0	3.2-3.5	Brittle; usually granular.
White_____	2.3	1.7	Tastes bitter and saline; in mineral waters.
Paler than color_	1.5-2.5	3.0	Deposits of secondary origin; with cobalt ores.
_____	5.0-5.5	7.2-7.5	Found with other tungsten ores.
White_____	4.0	3.0-3.3	Octahedral cleavage; brittle.
Brown to black_	5.5-6.5	5.2	Usually associated with zincite; sometimes magnetic.
Lead gray_____	3.0	7.4-7.6	Very brittle; cubic cleavage.
White_____	6.5-7.5	3.2-4.3	Often in complete dodecahedral crystals, in schists or limestone. Gem varieties.
Greenish white_	2.0-4.0	2.4	Amorphous; source of nickel; with serpentine, chromite.
Greenish white_	2.0-4.0	2.4	Similar to garnierite.
_____	2.0-3.5	2.4	Occurs under same conditions as bauxite.
Golden yellow__	2.8	15.6-19.3	Malleable and sectile. Does not tarnish.
Dark gray, iron black.	1.0-2.0	2.2	Soft; marks paper; feels greasy; often impure.
Yellow to red___	3.0-3.5	5.0	Usually occurs as coating on zinc minerals.
White_____	6.5-7.5	3.4-3.7	Often imbedded in mica and schists; limestones. Variety of garnet. See Garnet.
White to gray___	1.5-2.0	2.3	Earthy, fibrous, scaly, and crystals with perfect cleavage in one direction.
White_____	2.5	2.1-2.6	Natural table salt. Perfect cubical cleavage.
_____	1.0-2.0	2.0-2.2	Often occurs in veins of ore as secondary product.
Brown_____	5.3	4.7	Associated with other manganese minerals.
Red, brown_____	5.5-6.5	4.9-5.3	Becomes magnetic upon heating under reducing conditions.
Black_____	2.5-3.0	8.3-8.9	With chalcopyrite, pyrite, and sphalerite.
White_____	5.0-6.0	3.2	Crystals have 6-sided or diamond-shaped cross sections. Two perfect cleavages at angle of about 124°.
Yellowish brown.	5.0-5.5	7.2-7.5	Usually in bladed aggregates with rough, flat parting, in quartz.
White_____	2.0-2.5	3.6-3.8	Usually associated with other zinc ores.
Gray_____	5.0-6.0	3.5	Occurs in foliated or platy masses.
Brown_____	5.0-6.0	4.5-5.0	Magnetic; with pyrite, hornblende, feldspars.
_____	3.0-4.0	5.6-5.7	Usually in thin plates; rare.
Gray_____	6.7	22.7	With platinum and allied metals.
Gray_____	6.0-7.0	19.3-21.1	Rare metals alloy.
Grayish black__	2.0-3.0	5.5-6.0	Usually in parallel or divergent aggregates of narrow blades. Sometimes in hair- or needle-like crystals.

TABLE 1.—*Minerals and their*

Name	Formula	Percent metal	Color	Lustre
Jefferisite	Variable	Variable	Yellowish brown.	Pearly
Kainite	$MgSO_4 \cdot KCl \cdot 3H_2O$.	KCl—30.0%	White to red	Vitreous
Kaolinite	$H_4Al_2Si_2O_9$	Al_2O_3—39.5%	White, yellow	Pearly
Kermesite	Sb_2S_2O	Sb—75.3%	Cherry	Adamantine, metallic.
Kieserite	$MgSO_4 \cdot H_2O$	Mg—17.6%	White, yellow	Vitreous
Lepidolite	$KLi[Al(OHF)_2]Al(SiO_3)_3$.	Small amount of Li.	Red, lilac, white.	Pearly
Leucite	$KAl(SiO_3)_2$	K_2O—21.5% Al_2O_3—23.5%.	Gray	Vitreous, dull.
Limestones	Chiefly $CaCO_3$.	Ca—40%	Variable	Dull
Limonite	$2Fe_2O_3 \cdot 3H_2O$	Fe—59.9%	Brown, yellow.	Submetallic
Linnaeite	Co_3S_4	Co—58.0%	Steel gray	Metallic
Livingstonite	$HgS \cdot 2Sb_2S_3$	Hg—22.0%	Lead gray	Metallic
Magnesite	$MgCO_3$	Mg—28.9%	White to black.	Vitreous
Magnetite	$FeO \cdot Fe_2O_3$	Fe—72.4%	Iron black	Metallic, submetallic.
Malachite	$CuCO_3 \cdot Cu(OH)_2$.	Cu—57.5%	Green	Silky
Manganite	$Mn_2O_3 \cdot H_2O$	Mn—62.5%	Iron black, steel gray.	Metallic, submetallic.
Marble	Chiefly $CaCO_3$.	Ca—40%	Variable	Vitreous, earthy.
Marcasite	FeS_2	Fe—46.6%	Yellow	Metallic
Marmatite	(ZnFe)S, variable.	Zn—46.5% to 56.9%.	Yellow, brown, black.	Adamantine
Melaconite	CuO	Cu—79.9%	Black	Earthy, metallic.
Melilite	$Ca_{12}Al_4Si_9O_{36}$		White, yellow, green, brown.	Vitreous
Mercury	Hg	Hg—100%	Tin white	Metallic
Metacinnabarite	HgS	Hg—86.2%	Grayish black.	Metallic
Millerite	NiS	Ni—64.8%	Yellow	Metallic
Mimetite	$(PbCl)Pb_4As_3O_{12}$.	Pb—69.7%	Yellow to brown.	Resinous
Molybdenite	MoS_2	Mo—60%	Lead gray	Metallic
Molybdite	MoO_3	Mo—66.67%	Yellow	Adamantine, pearly.
Monazite	$(CelaDy)PO_4 \cdot ThSiO_4$.	ThO_2—9%	Yellow, brown.	Resinous
Mottramite	Variable	Variable	Black, yellow	Resinous
Muscovite	$H_2KAl_3(SiO_4)_3$	Variable	Yellowish white.	Vitreous, pearly.
Naumannite	$(Ag_2Pb)Se$	Ag—43.0%	Iron black	Metallic
Nephelite	$NaAlSiO_4$	No metal source.	White, yellow.	Vitreous, greasy.

72

Streak	Hardness	Specific gravity	Characteristics—occurrence
White	1.5	2.3	A mica with flexible but not elastic cleavage plates that puffs out greatly when heated.
	2.8	2.1	See Cyanite.
Same as color	2.0-2.5	2.6	Widespread; earthy odor; clay.
Brownish red	1.3	4.6	Occurs with stibnite.
	3.3	2.6	Often with gypsum and carnallite.
White	3.0	2.8-3.3	A mica with flexible, elastic cleavage plates. Usually in pegmatites.
White	5.5-6.0	2.5	Complete trapezohedral crystals in igneous rock.
White	3.0	2.7	Eliminate this heading since limestone is a rock, not a mineral.
Yellowish brown.	5.0-5.5	3.6-4.0	Massive, fibrous or porous; magnetic after fusing.
Blackish gray	5.5	4.8-5.0	Copper red tarnish; in gneiss with chalcopyrite.
Red	2.0	4.81	Resembles stibnite; fuses easily.
White	4.0-4.5	3.1	Effervesces vigorously in hot, concentrated muriatic acid.
Black	5.5-6.5	5.2	Strongly magnetic; many associations.
Green	3.5-4.0	4.0	Oxidized mineral. Effervesces vigorously in muriatic acid of any strength or temperature.
Brown	4.0	4.2-4.4	Hardness and streak are distinctive.
White, gray	3.0	2.7	Granular calcite. See calcite.
Grayish, brown, black.	6.0-6.5	4.9	Deposited near earth's surface. Often in tabular crystals in coxcomb-like groups.
Brownish	5.0	3.9-4.2	Closely allied with galena; common zinc ore.
	3.0-4.0	6.5	
White	5	2.9-3.1	Formed from magmas; common in portland cement.
		13.59	Liquid; rarely found in metallic state.
Black	3	7.7	Found in upper portions of mercury deposits.
Greenish black	3.0-3.7	5.3-5.7	Crusts with a radiating texture and hair- or needle-like crystals.
White	3.5	7.0-7.3	Often in crystals with 6-sided cross sections, which may taper.
Greenish gray	1.0-1.5	4.7-4.8	Feels greasy. Makes light greenish yellow mark on glazed paper.
	1.5	4.5	Occurs with molybdenite.
White	5.0-5.5	4.9-5.3	Rounded grains; with gold, chromite, iron.
Yellow	3	5.8	A vanadate of lead and copper.
White	2.0-2.5	2.8-3.0	Cleaves easily into very thin, elastic, flexible leaves.
Iron black	2.5	8	Malleable; in cubic crystals; selenide of silver and lead.
White	5.5-6.0	2.5-2.7	Widely distributed in igneous rocks; usually massive.

TABLE 1.—*Minerals and their*

Name	Formula	Percent metal	Color	Lustre
Niccolite	NiAs	Ni—44.1% As—55.9%.	Copper red	Metallic
Nitre	KNO₃	K—38.6% N—13.9%.	White	Vitreous
Olivine	(MgFe)₂SiO₄	No metal source.	Green	Vitreous
Opal	SiO₂·nH₂O	No metal source.	All colors	Greasy, vitreous.
Orpiment	As₂S₃	As—61%	Lemon yellow.	Resinous
Orthoclase	KAlSi₃O₈	Al₂O₃—18.4%	Red, gray, yellow, white.	Vitreous, dull.
Pentlandite	(FeNi)S	Fe—42.0% Ni—22.0%.	Yellow-bronze.	Metallic
Petzite	(AuAg)₂Te	Au—25.5% Ag—42%.	Gray to black.	Metallic
Phosphate rock	Ca₃(PO₄)₂	P₂O₅—31.1%	Gray	Dull
Platinum	Pt	Pt—100%	Tin white, steel white.	Metallic
Polianite	MnO₂	Mn—63.2%	Steel gray, iron gray.	Metallic
Polybasite	9Ag₂S·Sb₂S₃	Ag—75.6% Sb—9.4%.	Iron black	Metallic
Powellite	Ca(Mo·W)O₄	Variable	Greenish yellow.	Resinous
Proustite	3Ag₂S·As₂S₃	Ag—65.5%	Scarlet	Adamantine, dull.
Psilomelane	MnO₂·H₂O·K₂BaO₂.		Black	Submetallic, dull.
Pyrargyrite	3Ag₂S·Sb₂S₃	Ag—60% Sb—22.2%.	Black, reddish.	Adamantine, metallic.
Pyrite	FeS₂	Fe—46.7%	Brass yellow	Metallic
Pyrolusite	MnO₂	Mn—63.2%	Black, dark gray.	Metallic, dull
Pyromorphite	Pb₅Cl(PO₄)₃	Pb—76.4%	Yellow	Greasy, adamantine.
Pyrope	Mg₃Al₂(SiO₄)₃	No metal source.	Red	Vitreous, resinous.
Pyrophyllite	HAl(SiO₃)₂	Al₂O₃—28.3%	White, brown	Pearly, dull
Pyroxene	Ca(AlMg MnFe) (SiO₃)₂.	No metal source.	Green	Vitreous, dull.
Pyrrhotite	Fe₅S₆ to Fe₁₆S₁₇.	Fe—61.5% variable.	Brownish yellow.	Metallic
Quartz	SiO₂	Si—46.9%	Colorless, all colors.	Vitreous
Realgar	AsS	As—70.1%	Orange	Resinous
Rhodochrosite	MnCo₃	MnO—61.7%	Usually red	Vitreous, pearly.
Rhodonite	MnSiO₃	Mn—42.0%	Brownish red	Vitreous, dull
Roscoelite	H₈K(MgFe) (AlV)₄ (SiO₃)₁₂.	Variable	Brown	Pearly
Rutile	TiO₂	Ti—60%	Brown, red, black.	Adamantine, submetallic.
Scheelite	CaWO₄	W—63.9%	White yellowish.	Vitreous, adamantine.

Streak	Hardness	Specific gravity	Characteristics—occurrence
Brownish black.	5.0–5.5	7.3–7.7	Often found with a green coating; brittle; compact.
White	2	2.1	Tastes saline and cooling; salt petre.
White or yellowish.	6.5–7.0	3.3	Same as Chrysolite.
White	5.5–6.5	1.9–2.3	Very smooth, curving fracture.
Lemon yellow	1.5–2.0	3.5	Usually associated with realgar; seldom valuable.
White	6.0–6.5	2.5–2.6	Common, often pinkish igneous rock mineral with two smooth right angled cleavages.
Black	3.5–4.0	4.6–5.0	Associated with pyrrhotite, millerite, chalcopyrite, etc.
Gray	2.5	9.1	A rare but valuable ore of gold and silver; often tarnishes.
Gray	5	3.2	Occurs in massive deposits.
Shiny gray	4.5	17.0	Sometimes magnetic; with gold and chromite.
Black	6.3	4.9	Looks like pyrolusite, but harder and dryer; rare.
Black	2.0–3.0	6.0–6.2	With chalcopyrite, calcite, pyrargyrite, stephanite.
	3.5	4.5	Often associated with scheelite.
Scarlet	2.0–2.5	5.6	Usually associated with other silver ores.
Black, brownish black.	5.0–6.0	3.7–4.7	Either powdery (Wad) or has smooth, curving fracture.
Purplish red	2.5	5.8–5.9	Often associated with argentite and prousite.
Greenish, brownish black.	6.0–6.5	5.0	Often in crystals that are cubical or show prominently a form with 5-sided faces.
Black, bluish black.	1.0–2.5	4.8	Soils fingers; hardness and streak are distinctive.
White, yellowish white.	3.5–4.0	5.9–7.1	Alteration product of lead minerals. Occurs like mimetite.
White	6.5–7.6	3.7	Variety of garnet. See garnet.
White	1.0–2.0	2.8–2.9	Feels greasy or soapy.
White to green	5.0–6.0	3.3	Commonly in igneous rocks in square or 8-sided crystals.
Grayish black	3.5–4.6	4.6	Only magnetic sulphide and therefore distinctive.
White	7.0	2.65–2.66	Common in 6-sided prisms with pointed terminations. Gem varieties.
Orange	1.5–2.0	2.6	Usually associated with Orpiment; flexible.
White	3.5–4.5	3.5–3.6	Blackens on exposure. Effervesces vigorously in hot, concentrated muriatic acid.
White	·5.5–6.5	3.4–3.7	With calcite, Zincite, tetrahedrite.
	Soft	2.9	Vanadium mica in which vanadium replaced aluminum.
Light brown	6.0–6.5	4.2	Commonly in crystals with longitudinally grooved faces, or needle- or hair-like.
White	4.5–5.0	5.9–6.1	Weight, hardness, and uneven fracture are distinctive.

TABLE 1.—*Minerals and their*

Name	Formula	Percent metal	Color	Lustre
Senarmontite	Sb_2O_3	Sb—83.6%	Colorless, grayish.	Vitreous, dull.
Serpentine	$H_4Mg_3Si_2O_9$	Mg—43%	Green, blackish or yellow, white.	Wax-like, silky.
Siderite	$FeCO_3$	Fe—48.3%	Brown, gray	Vitreous, pearly, dull.
Silver	Ag	Ag—100%	Silver white	Metallic
Smaltite	$CoAs_2$	CO—28.2% As—71.8%.	Tin white, steel gray.	Metallic
Smithsonite	$ZnO \cdot CO_2$	Zn—52%	Green, gray, blue.	Vitreous, dull.
Soda nitre	$NaNO_3$		White, reddish brown; colorless.	Vitreous
Sperrylite	$PtAs_2$	Pt—56.6% As—43.4%.	Tin white	Metallic, brilliant.
Spessartite	Mn_3Al_2 $(SiO_4)_3$.	No metal source.	Purplish, red	Vitreous
Sphalerite	ZnS	Zn—67.1%	Brown, yellow, reddish, black.	Submetallic, resinous.
Spinel	$MgOAl_2O_3$	Al_2O_3—71.8% MgO—28.2%	Black, gray, brown, red.	Vitreous, dull.
Spodumene	$LiAl(SiO_3)_2$	Al_2O_3—27.4% Li_2O—8.4%.	White, grayish.	Vitreous, dull.
Stannite	$Cu_2S \cdot FeS \cdot SnS_2$.	Sn—27.5% Cu—29.5%.	Steel gray, iron black.	Metallic
Stephanite	$5Ag_2S \cdot Sb_2S_3$	Ag—68.5%	Iron black	Metallic
Stibnite	Sb_2S_3	Sb—71.8%	Lead gray	Metallic
Strontianite	$SrCO_3$	Sr—59.3%	Yellow to brown, green.	Vitreous, greasy.
Sulfur	S	S—100%	Yellow	Greasy, adamantine.
Sylvanite	$(AuAg)Te_2$	Au—24.5% Ag—13.4%.	White to steel gray.	Metallic
Sylvite	KCl	K—52.4%	White, yellowish red.	Vitreous
Talc	$H_2Mg_3(SiO_3)_4$	Mg—19.2% Si—29.6%.	Green to white.	Pearly
Tantalite	$FeTa_2O_6$	Variable Ta_2O_5—65.6%.	Iron black	Submetallic, greasy, dull.
Tennantite	$Cu \cdot As_2S_7$ variable.	Cu—57.5% variable.	Steel gray, iron black.	Metallic
Tenorite	CuO	Cu—79.9%	Black	Metallic
Tephroite	Mn_2SiO_4	No metal source.	Red, ash gray	Vitreous
Tetradymite	$Bi_2(TeS)$	Variable	Pale steel gray	Metallic
Tetrahedrite	$4Cu_2S \cdot Sb_2S_3$	Cu—52.1% Sb—24.8%.	Gray to black	Metallic
Titanite	$CaTiSiO_5$	TiO_2—40.8%	Brown, gray, yellow, green.	Adamantine

76

Streak	Hardness	Specific gravity	Characteristics—occurrence
White	2	5.3	Formed by oxidation of stibnite.
White	4.0	2.5–2.6	Feels smooth and sometimes slightly greasy.
White to yellow	3.5–4.0	3.9	Magnetic after heating. Effervesces vigorously in hot, concentrated muriatic acid.
Silver-white	2.8	10.5	Malleable and sectile. Tarnishes quickly.
Grayish black	5.5–6.0	5.7–6.8	Granular or in crystals like pyrite. Often with erythite. See erythite.
White, grayish	5.0	4.3–4.5	Effervesces vigorously in any strength or temperature or muriatic acid except cold dilute.
White	1.8	2.3	Taste-cooling; incrustations in beds; massive.
Black	6.5	10.6	Found with gold-quartz, covellite, limonite.
	6.5–7.5	4.0–4.3	A form of garnet.
Light brown, yellow.	3.5–4.0	3.9–4.1	Cleaves smoothly in six directions at angles of 60°, 90°, and 120°.
White to gray	8.0	3.5–4.1	Massive or in octahedral crystals. Gem varieties.
White	6.5–7.0	3.1–3.2	Occurs usually in platy masses or chunky crystals, sometimes huge. Gem varieties.
Blackish	4.0	4.5	Has appearance of bronze.
Iron black	2.0–2.5	6.2–6.3	Associated with other silver ores.
Lead gray, black.	2.0	4.5–4.6	Cleavage surfaces marked transversely with parallel lines.
White to gray	3.5–4.0	3.7	Effervesces vigorously in dilute cold, but not in concentrated cold, muriatic acid. Effervescing fragment colors alcohol flame red.
Pale yellow	2.0	2.0	Burns with a characteristic odor.
Same as color	1.5–2.0	7.9–8.3	Occurs often in small, bladed or prismatic crystals.
White	2.0	1.98	Taste—saline; soluble; bitter.
White	1.0–1.5	2.7–2.8	Common; feels greasy; extensive beds.
Reddish brown	6.3	5.3–7.3	Iron and manganese content variable; with columbite.
Black, reddish brown	3.0–4.5	4.4–4.5	Occurs granular massive or in tetrahedral crystals.
	3.0	5.8–6.3	Sublimation product in volcanic regions.
	6.5–7.0	4.0–4.1	Rarely in small crystals; like chrysolite.
	1.8	7.4	Soils paper; found in gold-quartz and igneous rocks.
Black	3.0–4.5	4.4–5.1	Like tennanite but has a darker streak—not reddish.
White	5.0–5.5	3.4–3.6	Occurs in platy massive or in wedge-shaped crystals.

TABLE 1.—*Minerals and their*

Name	Formula	Percent metal	Color	Lustre
Topaz	$(AlF)_2SiO_4$	No metal source.	Many	Vitreous
Tourmaline	$[(NaLiK)_6 (MgFeCa)_3 (AlCrFe)_2 B_2SiO_5]$	No metal source.	Black, brown, and many others.	Vitreous to resinous.
Tremolite	$CaMg_3(SiO_3)_4$	No metal source.	White to dark gray.	Silky
Triphyllite	$LiFePO_4$	Li—4.4%	Greenish gray, bluish gray.	Vitreous, resinous.
Ullmannite	$NiSbS$	Ni—27.6% Sb—57.3%.	Steel gray to white.	Metallic
Uraninite	UO_3, UO_2 variable.	Radium source.	Gray, green, brown.	Submetallic to greasy.
Uvarovite	$Ca_3Cr_2(SiO_4)_3$	No metal source.	Green	Vitreous
Valentinite	Sb_2O_3	Sb—83.5%	White	Adamantine to pearly.
Vanadinite	$(PbCl)Pb_4 (VO_4)_3$.	Variable.	Red, brown, yellow.	Resinous
Vermiculite	$3MgO. (FeAl)_2O_3 3SiO_2$.	Variable	Grayish	Talc-like
Willemite	Zn_2SiO_4	Zn—58.5%	Green, yellow, brown.	Vitreous, dull.
Witherite	$BaCO_3$	BaO—77.7%	Yellow, brown.	Vitreous, pearly.
Wolframite	$(FeMn)WO_4$	W—51.3%	Gray, brown, black.	Submetallic
Wulfenite	$PbMoO_4$	Pb—56.4% Mo—26.2%.	Yellow, grayish.	Resinous, adamantine.
Zaratite	$NiCO_3. 2Ni(OH)_2 4H_2O$.	Ni—46.8%	Green	Vitreous
Zincite	ZnO	Zn—80.3%	Red, yellow	Sub-adamantine.
Zircon	$ZrSiO_4$	ZrO—67.2%	Yellow, gray	Adamantine

Streak	Hardness	Specific gravity	Characteristics—occurrence
------------------	8.0	3.4–3.6	Often in prismatic crystals with diamond-shaped cross sections and a perfect transverse cleavage. Gem varieties.
White _____	7.0–7.5	3.0–3.2	Usually in prismatic crystals with spherical triangular cross sections. Gem varieties.
White _____	5.0–6.0	2.9–3.4	Perfect cleavages in two directions at an angle of about 124°.
White to grayish white.	4.8	3.5	A phosphate or iron, manganese and lithium.
Grayish_____	5.3	6.4	With galena and chalcopyrite.
Black, gray, green.	5.5	9.0–9.7	Of primary and secondary origin; no definite formula.
White_____	6.5–7.5	3.5	A form of garnet.
White _____	2.5–3.0	5.6	An oxidized mineral.
White or yellow.	2.7–3.0	6.6–7.1	Like mimetite, but crystals usually very sharp and do not taper.
Uncolored _____	1.5	2.7	Becomes worm-like threads upon heating—exfoliates.
White or grayish.	5.5	3.9–4.2	Massive to granular; valuable zinc ore.
White_____	3.4	4.4	Reacts like strontianite in muriatic acid, but effervescing fragment colors alcohol flame light yellowish green.
Reddish-brown.	5.0–5.5	7.2–7.5	Differs from huebnerite in streak.
White_____	3.0	6.8	In square crystals, usually tabular with beveled edges.
Light green_____	3	2.6	Emerald nickel; amorphous.
Orange yellow__	4.0–4.5	5.4–5.7	Associated with other zinc ores.
Colorless_____	7.5	4.2–4.7	In sharp crystals with square cross sections and as pebbles. Gem varieties.

COMMERCIALLY IMPORTANT ORES*

In the following table, the figures after each name of an ore indicate the percentage of the element specified which the pure mineral contains. When this is variable or is merely mechanically included, an interrogation mark takes the place of the above-mentioned figure.

Important ores are in **heavy face type**, less common species are in lighter type, and minerals which are only occasionally mined and treated for the element specified are in *italics*.

Each group is arranged in the order of decreasing importance.

ALUMINUM: Bauxite (39.2), **Cryolite** (12.8).

ANTIMONY: Stibnite (71.8).

ARSENIC: Arsenopyrite (46), **Smaltite** (71.8), **Cobaltite** (45.2), Nicolite (?), Enargite (19.1).

BARIUM: Witherite (65).

BISMUTH: Bismuthinite (40.6).

CHROMIUM: Chromite (46.2).

COBALT: Smaltite (?), **Cobaltite** (35.5), *Arsenopyrite (?)*.

COPPER: Native Copper (95), **Chalcopyrite** (34.5), **Bornite** (55.5), **Cuprite** (88.8), Malachite (57.5), Chalcocite (79.8), Enargite (48.3), Tetrahedrite (?), Azurite (55.4), *Covellite (66.4)*, *Chrysocolla (45.2)*, *Atacamite (62.4)*, *Tenorite (79.9)*.

GOLD: Native Gold (99.8), **Pyrite** (?), **Sylvanite** (24.5), **Calavarite** (39.5), **Chalcopyrite** (?), Hessite (?), Petzite (25.5), Galenite (?), *Arsenopyrite (?)*, *Stibnite (?)*.

IRON: Hematite (70), **Limonite** (59.8), **Magnetite** (72.4), Siderite (48.4), Goethite (62.9), *Pyrite (46.7)*.

LEAD: Galena (86.6), **Cerussite** (77.7), **Anglesite** (73.6), **Pyromorphite** (76.4), Mimetite (69.7), Vanadinite (73.2), Wulfenite (56.5), *Tetrahedrite (?)*.

LITHIUM: Amblygonite (4.7), Spodumene (3.7).

MAGNESIUM: Magnesite (28.6), Dolomite (21.9).

MANGANESE: Pyrolusite (63.2), **Psilomelane** (?), Manganite (62.4).

MERCURY: Cinnabar (86.2), *Native Mercury (99)*.

MOLYBDENUM: Molybdenite (60).

NICKEL: Garnierite (?), **Pyrrhotite** (?), Millerite (64.4), Niccolite (43.9), *Chalcopyrite (?)*, *Arsenopyrite (?)*.

PLATINUM: Native Platinum (86.5).

SILVER: Galenite (?), **Cerargyrite** (75.3), **Pyrargyrite** (59.9), **Proustite** (65.4), **Argentite** (87.1), Tetrahedrite (?), Native Silver (95), Native Gold (?), Native Copper (?), Hessite (63), Petzite (43), Stephanite (68.5), *Pyrite (?)*, *Chalcopyrite (?)*, *Jamesonite (?)*, *Stibnite (?)*, *Cerussite (?)*, *Polybasite (75.6)*.

STRONTIUM: Strontianite (56.8), Celestite (45.7).

SULPHUR: Pyrite (53.3), **Native Sulphur** (100), Pyrrhotite (?).

THORIUM: Monazite (?).

TIN: Cassiterite (78.6).

TITANIUM: Rutile (59.9).

TUNGSTEN: Wolframite (60.7), **Huebnerite** (60.7), *Scheelite (63.9)*.

*Handbook of Minerals, by C. Montague Butler, E.M.

URANIUM: Uraninite (?), **Carnotite** (?), Autunite (51.9), Torbernite (50.8), *Samarskite (?)*.
VANADIUM: Vanadinite (10.8), **Carnotite** (?).
ZINC: Sphalerite (67), **Smithsonite** (52), **Calamine** (54.1), Zincite (80.8), *Franklinite (?)*, *Willemite (58.4)*.

Note: In the foregoing table, Marcasite is included under Pyrite, and Tennantite under Tetrahedrite.

FLUORESCENT MINERALS

The fluorescence of certain minerals using a Mineralight* is an important characteristic in the identification of minerals.

Scheelite is extremely difficult to identify without the ultra-violet Mineralight. The fluorescent colors which indicate the presence of scheelite are blue, blue-white, cream, and a golden yellow. Occasionally a form of calcium carbonate will fluoresce and resemble scheelite very closely. For this reason the mineral should also be checked against its physical properties to confirm the identification. Scheelite never phosphoresces.

Another valuable ore which fluoresces is hydrozincite. This is frequently associated with smithsonite. It always fluoresces a soft blue. It can easily be distinguished from scheelite because it is a light weight rock, the ore is soft, and the fluorescence is usually (but not always) in the form of a heavy coating.

Black sand very often contains small bright orange fluorescent grains. These are zircon. They are brighter orange than scheelite and usually appear like grains of sand.

The most vivid and beautiful fluorescent rocks in the world are the willemite and calcite rocks of New Jersey. The willemite has a bright green fluorescence and the calcite is frequently a gorgeous red.

Mineral	*Fluorescence*
Autunite	Yellow green.
Beta—Uranopilite	Yellow green.
Beta—Uranotil	Yellowish.
Chalcolite	Yellow green.
Gummite (variable)	Violet.
Johannite (variable)	Yellow green.
Meta—Tobernite	Yellowish blue.
Schroeckingerite (Dakeite)	Green.
Tobernite	Yellow green.
Uranocircite	Yellow green.
Uraniferous hyalite	Yellow green.
Uranophane	Yellow green.
Uranopilite	Yellow green.
Uranospathite	Yellow green.
Uranothallite	Green.
Uranotil	Yellowish.
Zippeite	Yellowish.

A complete list of fluorescent minerals and other information relative to fluorescence will be supplied upon request. Ask for Deco Bulletin No. G3–B15.

*See page 25 for more information on Mineralight.

TABLE 2.—*Oxidation behavior of common metallic compounds*

Metal	Primary compounds	Important solutions formed	Oxidized compounds formed in oxidation zone	Transported out of oxidation zone
Iron	Sulphides	Ferric sulphate, colloidal.	Hematite, limonite, basic sulphates.	Yes.
	Carbonates	---------------	Limonite, ferric hydroxide.	No.
	Oxides	---------------	Hydrous ferric oxide.	No.
Copper	Sulphides, etc	Copper sulphate.	Carbonates, oxides, native, silicate.	Yes.
Lead	Sulphides, etc	---------------	Sulphate, carbonate.	No.
Zinc	Sulphide	Zinc sulphate	Carbonate, silicate, oxide.	Yes.
Tin	Oxide, sulphide	(?)	Oxide	No.
Aluminum	Silicates	Colloidal	Oxides, silicates	No.
Silver	Sulphides, etc	Silver sulphate, colloidal.	Chloride, iodide, bromide, native.	Yes.
Gold	Au, tellurides	Acid ferric sulphate +Cl.	Native	Yes.
Nickel	Sulphide	Nickel sulphate	Nickel arsenide, nickel silicate.	Yes(?).
Cobalt	Sulphide, etc	Cobalt sulphate	Cobalt oxides	Yes.
Molybdenum.	Sulphide, etc	---------------	Oxides	(?).
Chromium	Oxide	(?)	Oxides	No.
Tungsten	Oxides	---------------	Oxides	No.
Manganese	Oxides, carbonates.	Mn sulphate	Oxides	Yes; no.
		Mn bicarbonate, colloidal.	Sulphate	Yes.
			Oxides	No; yes.
Vanadium	Sulphide, oxide	Sulphate	Oxides	(?).
Mercury	Sulphide	---------------	Oxides, chloride	No.
Arsenic	As, arsenides	(?)	Arsenate, oxides, sulphides.	Yes.
Antimony	Sb, antimonides	(?)	Oxides	Yes.
Bismuth	Sulphides	---------------	Oxides	No.
Cadmium	Sulphide	Sulphate	Carbonate	No.
Platinum	Pt, sulphide	(?)	Arsenide	Yes.
Uranium	Oxides	(?)	Oxides, etc	(?).

TABLE 3.—*Common hypogene, supergene, and oxidized ore minerals*

Metal	Minerals generally hypogene	Minerals generally supergene	Minerals generally originating in oxidized zone
Copper	Chalcopyrite Bornite Enargite* Tetrahedrite* Tennantite*	Chalcocite Sooty chalcocite* Covellite	Native copper. Malachite.* Brochantite.* Antlerite.* Atacamite.* Azurite.* Chrysocolla.* Cuprite.* Tenorite.*
Silver	Tetrahedrite* Tennantite*	Native silver Argentite Pyrargyrite Proustite Stephanite Polybasite Pearcite	Cerargyrite.* Embolite.* Bromyrite.*
Gold	Native gold Gold tellurides*	Native gold	Native gold(?).
Zinc	Sphalerite Willemite*	Wurtzite	Smithsonite.* Hemimorphite.* Hydrozincite.*
Lead	Galena*		Cerussite.* Anglesite.* Pyromorphite.* Leadhillite.
Iron	Pyrite* Marcasite Pyrrhotite* Arsenopyrite* Magnetite* Hematite Specularite* Siderite	Marcasite	Goethite.* Iron sulphates.* Hematite.
Manganese	Rhodochrosite* Rhodonite* Manganite(?) Alabandite*		Psilomelane. Pyrolusite. Braunite.
Nickel	Millerite Pentlandite* Niccolite*	Bravoite(?)	Garnierite.*

*Always hypogene or supergene or oxidized according to the column in which they are placed.

TABLE 4.—*Common gangue minerals*

Class	Name	Composition	Pri-mary	Super-gene
Oxides	Quartz	SiO_2	X	X
	Other silica	SiO_2	X	X
	Bauxite, etc	$Al_2O_3 \cdot 2H_2O$		X
	Limonite	$Fe_2O_3 \cdot H_2O$		X
Carbonates	Calcite	$CaCO_3$	X	X
	Dolomite	$(Ca,Mg)CO_3$	X	X
	Siderite	$FeCO_3$	X	X
	Rhodochrosite	$MnCO_3$	X	
Sulphates	Barite	$BaSO_4$	X	
	Gypsum	$CaSO_4 + 2H_2O$		X
Silicates	Feldspar		X	
	Garnet		X	
	Rhodonite	$MnSiO_3$	X	
	Chlorite		X	
	Clay minerals		X	X
Miscellaneous	Rock matter		X	
	Fluorite	CaF_2	X	
	Apatite	$(CaF)Ca_4(PO_4)_3$	X	
	Pyrite	FeS_2	X	X
	Marcasite	FeS_2	X	
	Pyrrhotite	Fe_7S_8	X	
	Arsenopyrite	$FeAsS$	X	

84

TABLE 5.—*Common ore minerals*

Metal	Ore mineral	Composition	Percent metal	Primary	Supergene
Gold	Native gold	Au	100	X	X
	Calaverite	$AuTe_2$	39	X	
	Sylvanite	$(Au,Ag)Te_2$	------	X	
Silver	Native silver	Ag	100	X	X
	Argentite	Ag_2S	87	X	X
	Cerargyrite	$AgCl$	75	------	X
Iron	Magnetite	$FeO \cdot Fe_2O_3$	72	X	
	Hematite	Fe_2O_3	70	X	X
	"Limonite"	$Fe_2O_3 \cdot H_2O$	60	------	X
	Siderite	$FeCO_3$	48	X	X
Copper	Native copper	Cu	100	X	X
	Bornite	Cu_5FeS_4	63	X	X
	Brochantite	$CuSO_4 \cdot 3Cu(OH)_2$	62	------	X
	Chalcocite	Cu_2S	80	X	X
	Chalcopyrite	$CuFeS_2$	34	X	X
	Covellite	CuS	66	X	X
	Cuprite	Cu_2O	89	------	X
	Enargite	$3Cu_2S \cdot As_2S_5$	48	X	
	Malachite	$CuCO_3 \cdot Cu(OH)_2$	57	------	X
	Azurite	$2CuCO_3 \cdot Cu(OH)_2$	55	------	X
	Chrysocolla	$CuSiO_3 \cdot 2H_2O$	36	------	X
Lead	Galena	PbS	86	X	
	Cerussite	$PbCO_3$	77	------	X
	Anglesite	$PbSO_4$	68	------	X
Zinc	Sphalerite	ZnS	67	X	
	Smithsonite	$ZnCO_3$	52	------	X
	Calamine	$H_2Zn_2SiO_5$	54	------	X
	Zincite	ZnO	80	X	
Tin	Cassiterite	SnO_2	78	X	(?)
	Stannite	$Cu_2S \cdot FeS \cdot SnS_2$	27	X	(?)
Nickel	Pentlandite	$(Fe,Ni)S$	22	X	
	Garnierite	$H_2(Ni,Mg)SiO_3 \cdot H_2O$	------		X
Chromium	Chromite	$FeO \cdot Cr_2O_3$	68	X	
Manganese	Pyrolusite	MnO_2	63	------	X
	Psilomelane	$Mn_2O_3 \cdot xH_2O$	45	------	X
	Braunite	$3Mn_2O_3 \cdot MnSiO_3$	69	(?)	X
	Manganite	$Mn_2O_3 \cdot H_2O$	62	------	X
Aluminum	Bauxite	$Al_2O_3 \cdot 2H_2O$	39	------	X
Antimony	Stibnite	Sb_2S_3	71	X	
Bismuth	Bismuthinite	Bi_2S_3	81	X	X
Cobalt	Smaltite	$CoAs_2$	28	X	
	Cobaltite	$CoAsS$	35	X	
Mercury	Cinnabar	HgS	86	X	
Molybdenum	Molybdenite	MoS_2	60	X	
	Wulfenite	$PbMoO_4$	39	------	X
Tungsten	Wolframite	$(Fe,Mn)WO_4$	76	X	
	Huebnerite	$MnWO_4$	76	X	
	Scheelite	$CaWO_4$	80	X	

TABLE 6.—*Important gemstones*

Stone	Chief constituents	Common color	Hardness	Chief source
PRECIOUS STONES				
Diamond	C	Colorless to straw.	10	South-central Africa, Brazil.
Emerald	Be, Al, Si, O	Green	7.5–8.5	Colombia, Egypt.
Ruby	Al, O	Red	9	Burma, Ceylon.
Sapphire	Al, O	Blue	9	Ceylon, Burma, Siam.
Precious opal	Si, O	Variegated	5.5–6.5	Australia, Hungary, Mexico.
SEMIPRECIOUS STONES				
Amethyst	Si, O	Purple	7	India, Persia, Brazil.
Beryl	Be, Al, Si, O	Various	7.5–8.5	U.S., Africa.
Benitoite	Ba, Ti, Si	Deep blue	6.5	
Chrysoberyl	Be, Al, O	Green, yellow	8.5	
Feldspar	K, Na, Ca, Al, Si.	Various	6	
Garnet	Al, Fe, Mg, Si.	Red, green	6.5–7.5	Arizona, South Africa, Urals.
Jade—nephrite	Ca, Mg, Fe, Si.	Green to white.	5.5	Burma.
Jade—jadeite	Na, Al, Si	Green	6.5	Burma, China.
Kunzite	Li, Al, Si	Lilac	6.5–7	California, Madagascar.
Lapis lazuli	Na, Al, S, Si	Dark blue	5–5.5	India, Greece, California, Siberia.
Peridot	Mg, Fe, Si	Green	6.5–7	Levant, Egypt, Burma.
Quartz	Si, O	Various	7	Worldwide.
Spinel	Mg, Al	Reddish	8	Ceylon, Burma, Siam.
Topaz	Al, Fe, Si	Yellow	8	Brazil, Ceylon, Montana.
Tourmaline	Bo, Si, +	Green, pink	7–7.5	Urals, Madagascar, California, Maine.
Turquois	Al, P, O, H	Blue	5–6	New Mexico, Persia, Turkestan, Egypt.
Zircon	Zr, Si	Red, orange	7.5	Ceylon, Siam.

TABLE 7.—*Chemical composition and physical characteristics of the chief heavy minerals found in placer gravels*

Mineral	Chemical formula	Color	Specific gravity	Hardness	Remarks
Gold	$Au(+Ag)$	Gold-yellow	15.6-19.3	2.5	Very malleable and ductile.
Magnetite	Fe_3O_4	Iron-black	5.2	5.5-6.5	Shiny grains; strongly magnetic.
Ilmenite	$(Mg,Fe)TiO_3$	do	4.5-5	5-6	Only faintly magnetic; moves compass needle slightly.
Garnet	$R''_3R'''_2(SiO_4)_3$*	Red, brown, various	3.8	6.5-7.5	Vitreous luster; usually in rounded crystals (dodecahedron).
Zircon	$ZrSiO_4$	Brown, pale yellow, or colorless.	4.7	7.5	Adamantine luster.
Hematite	Fe_2O_3	Dark steel-gray to iron-black.	4.9-5.3	5.5-6.5	Particles usually smooth, rounded, often red-coated.
Chromite	$FeCr_2O_4$	Iron-black to brown-black.	4.1-4.9	5.5	Sometimes feebly magnetic; brown streak.
Olivine	$(Mg,Fe)_2SiO_4$	Olive-green	3.3-3.4	6.5-7	Good cleavage; vitreous luster; clear to translucent.
Epidote	$HCa_2(Al,Fe)_3 Si_3O_{13}$.	Pistachio-green	3.2-3.5	6-7	Distinct cleavage.
Pyrite	FeS_2	Pale brass-yellow	4.9-5.1	6-6.5	Usually cubic grains; brittle; metallic luster.
Monazite	$(Ce,La,Di)PO_4+ ThO_2$.	Yellow	4.9-5.3	5-5.5	Resinous or greasy luster; usually in rounded grains.
Limonite	$2Fe_2O.3H_2O$	Dark brown	3.6-4.0	5-5.5	Yellow-brown streak.
Rutile	TiO_2	Red-brown to red	4.2	6-6.5	Distinct cleavage; metallic-adamantine luster.
Platinum	Pt (usually also Fe. Ir,Os),	Whitish steel	16.5-18	4-4.5	Malleable; sometimes scales and grains.
Iridium	Ir (also Pt, etc.)	Silver-white, yellow tarnish.	22.6-22.8	6-7	Generally in angular grains; no cleavage.
Iridosmine	Ir,Os	Tin-white to light steel-gray.	19.3-21.1	6-7	Usually flat grains; slightly malleable to brittle; good cleavage.
Wolframite	$(Fe,Mn)WO_4$	Black, dark gray	7.2-7.5	5-5.5	Submetallic luster; good cleavage in one plane.
Cinnabar	HgS	Red	8-8.2	2-2.5	Scarlet streak.

87

TABLE 7.—*Chemical composition and physical characteristics of the chief heavy materials found in placer gravels*—Continued

Mineral	Chemical formula	Color	Specific gravity	Hardness	Remarks
Sheelite	CaWO₄	White, pale yellow, brown, or gray.	5.9–6.1	4.5–5	Adamantine, greasy luster; translucent.
Cassiterite	SnO₂	Brown or black	6.8–7.1	6–7	Brittle; rounded grains.
Corundum Sapphire Ruby	} Al₂O₃	Blue, red, yellow, brown.	3.9–4.1	9	Adamantine to vitreous luster.
Diamond	C	White, colorless, pale	3.5	10	Adamantine or greasy luster.
Mercury	Hg	Tin-white	13.6		Small opaque fluid; silvery globules.
Amalgam	Hg, Ag, Au	Silver-white	13–14		Brittle to malleable; rubs silvery coat on copper.
Silver	Ag	Silver-white	10.1–11.1	2.5–3	Malleable and ductile; tarnishes black.
Copper	Cu	Copper-red	8.8–8.9	2.5–3	Ductile; malleable.
Bismuth	Bi	Silver-white	9.8	2.5	Sectile; brittle; metallic luster.
Columbite-tantalite	(Fe,Mn)(Nb,Ta)₂O₆	Iron-black to gray or brown-black.	5.3–7.3	6	Brilliant to submetallic luster, often irridescent; brittle; good cleavage.
Quartz	SiO₂	Colorless	2.6	7	No cleavage; vitreous to greasy luster.
Feldspars	Silicates of K, Na, Ca, Al, etc.	Colorless, white, pale yellow, or pink.	2.5–2.7	6–6.5	Good cleavage; vitreous luster.
Galena	PbS	Lead-gray	7.4–7.6	2.5–2.7	Metallic luster; lead-gray streak; perfect cubic cleavage; friable.
Cerussite	PbCO₃	Colorless or white	6.5	3–3.5	Adamantine luster.

*R″ = Ca, Mg, Mn, or Fe; R‴ = Fe, Al, or Cr.

88

TABLE 8.—*Elements that can be determined spectrographically*

Element	Granite (percent)	Diabase (percent)	Element	Granite (percent)	Diabase (percent)
BeO	0.001	n.d.[*]	CoO	0.0005	0.004
ZrO_2	.1	0.003	Cr_2O_3	.001	.03
La_2O_3	.005	n.d.[*]	U_2O_3	.0005	.01
Y_2O_3	.01	n.d.[*]	Sc_2O_3	.001	.003
Nd_2O_3	.003	n.d.[*]	CuO	.005	.01
BaO	.2	.005	GeO_2	.0005	n.d.[*]
SrO	.05	.10	Tl_2O	.0003	n.d.[*]
Rb_2O	.05	n.d.[*]	Nb_2O_5[**]	.01	n.d.[*]
Cs_2O	.002	n.d.[*]	Ag_2O	n.d.[*]	.0002
Li_2O	.01	.002	SnO_2	.01	n.d.[*]
Ga_2O_3	.002	.002	B_2O_3	.003	n.d.[*]
PbO	.003	.004	F	.2	.02
NiO	.0005	.005			

[*]Usually not detectable.
[**]Sometimes reported as "Cb" for columbium. Columbium and niobium are alternate names for the same element, though niobium is the preferred name.

89

TABLE 9.—*Minerals of hydrothermal deposits*

Ore minerals	Hypothermal	Mesothermal	Leptothermal	Epithermal	Gangue and rock-alteration minerals	Hypothermal	Mesothermal	Leptothermal	Epithermal
Magnetite	X	(x)			Garnet	X	X		
Specularite	X	(x)	(x)		Pyroxene	X	X		
Pyrrhotite	X	(x)			Amphibole	X	X		
Cassiterite	X	(x)			Forsterite	X	X		
Arsenopyrite	X	X	(x)	(x)	Ilvaite	X	X		
Bismuthinite	X	---	---	(x)	Vesuvianite	X	X		
Molybdenite	X	X	(x)	(x)	Anorthite	X	X		
Bornite	X	X	(x)	(x)	Wallastonite	X	X		
Gold (native)	X	X	X	X	Axinite	X	X		
Pyrite	X	X	X	X	Scapolite	X	X		
Sphalerite	X	X	X	X	Biotite	X	X		
Galena	X	X	X	X	Muscovite	X	X		
Chalcopyrite	X	X	X	X	Topaz	X	X		
Enargite (Famatinite)	---	X	X	(x)	Tourmaline	X	(x)		
Chalcocite	---	X	(x)	(x)	Albite	X	X	X	X
Jamesonite	---	(x)	X	X	Epidote	X	X	X	X
Bournonite	(x)	X	X		Quartz	X	X	X	X
Boulangerite	(?)	---	X	(?)	Sericite	(x)	X	X	(x)
Silver (native)	---	---	X		Chlorite (high iron)	X	(x)		
Cobaltite	---	---	X		Chlorite (low iron)	---	(x)	X	X
Niccolite	---	---	X		Carbonates	X	X	X	X
Smaltite	---	---	X		Fluorite	X	(x)	(x)	X
Ruby Silvers	---	---	(x)	X	Rhodonite	X	X	(x)	
Polybasite	---	---	(x)	X	Siderite	---	X	X	X
Pearceite	---	---	(x)	X	Rhodochrosite	---	---	X	X
Stephanite	---	---	(x)	X	Barite	---	(x)	X	X
Marcasite	---	---	---	X	Dickite	---	(x)	X	X
Stibnite	---	(x)	(x)	X	Adularia	---	---	(x)	X
Bismuth (native)	---	---	---	X	Alunite	---	---	---	X
Argentite	---	---	(x)	X					
Cinnabar	---	(x)	---	X					
Selenides	---	---	---	X					
Realgar	---	---	---	X					
Orpiment	---	---	---	X					

NOTE. X—Common or characteristic.   (x)—Sparse or occasional.   (?)—Doubtful occurrence (contingent on classification of specific deposits).

90

TABLE 10.—*Grain size—Wentworth classification*

Size limits	Pieces	Aggregates	Cemented rock
256 mm____	Rounded-boulders. Angular-blocks.	Gravel.	Conglomerate. Breccia.
64 mm____	Rounded-cobbles. Angular-blocks.	Gravel.	Conglomerate. Breccia.
4 mm_____	Rounded-pebbles. Angular-blocks.	Gravel.	Conglomerate. Breccia.
2 mm____	Rounded} Angular }granules.	Gravel (grit).	Conglomerate (grit).
1 mm____	Very coarse sand grains.	Very coarse sand.	Very coarse sandstone.
½ mm____	Coarse sand grains.	Coarse sand.	Coarse sandstone
¼ mm____	Medium sand grains.	Medium sand.	Medium sandstone.
⅛ mm____	Fine sand grains.	Fine sand.	Fine sandstone.
¹⁄₁₆ mm___	Very fine sand grains.	Very fine sand.	Very fine sandstone.
¹⁄₂₅₆ mm__	Silt particles.	Silt.	Siltstone }{shale
	Clay particles including colloids.	Clay.	Claystone }{mudstone argillite

91

TABLE 11.—*Data on metals and their ores*

Metal	Unit of measurement	Tenor of Ore		Common associates	Commercial unit	Specifications	Prices 1959
		Low grade	Average				
Gold	$/ton	2.00	5-12	Ag	oz troy		$35.00
Silver	oz/ton	10	12-30	Ag, Pb	oz troy		0.912
Platinum	dwt/ton	3	4-8	Pt group	oz troy		78.750
Iron	% Fe	30	40-60		l.t. ore[3]		11.611
Copper	% Cu	0.8	1.5-5	Au, Ag	lb cu		0.318
Lead	% Pb	5	6-10	Zn, Ag	lb Pb		28.892
Zinc	% Zn	3	10-30	Pb	lb Zn		0.1145
Tin	% Sn	1	1, 5-5		lb Sn		1.02
Nickel	% Ni	1.5	2-3	Cu, Pb	lb Ni	Cathode	0.740
Aluminum	% Al_2O_3	30	55-65		lb Al	99+%	0.281
Antimony	% Sb	40	50-60	Ag	lb Sb	Bulk f.o.b. Laredo	0.290
Bismuth	% Bi	By-product	40-60	W	lb Bi		2.25
Beryllium	% BeO	10	10-12		Stu BeO[1]	10% BeO	32.50
Arsenic	% As_2O_3	By-product			lb As_2O_3		0.04
Cobalt	% Co	5	8-10	Ag, Cu	lb Co	97-99%	1.75
Chromium	% Cr_2O_3	32	35-50		l.t. Cr_2O_3[3]	44% N.R.[2] dry	18.75
					l.t. Cr_2O_3[3]	48% 3 to 1 dry	34.00
Cadmium	% Cd	By-product		Fe	lb Cd	Small lots	1.50
Manganese	% Mn	35	45-55		ltu/ton[4]	48% Mn	0.90
Mercury	% Hg	0.5	1-3		Flask-76 lb		227.48
Molybdenum	% MoS_2	0.4	1-3		lb Mo	90%	1.25
Titanium	% TiO_2	3	4-8	Fe	lb TiO_2	99.3+% max 3% Fe	1.60
Tungsten	% WO_3		60-70		Stu WO_3[1]	60%	22.00-24.00
Vanadium	% V_2O_5	2	3-8		lb V_2O_5	F.o.b. mines	0.31

[1] Short ton unit.
[2] No ratio.
[3] Long ton.
[4] Long ton unit.

92

TABLE 12.—*Common and technical names*

Common name	Technical name	Common name	Technical name
Agatized wood	Agate.	Desmine	Stilbite.
Alum stone	Alunite.	Dichroite	Iolite.
Amber mica	Phlogopite.	Electric calamine	Calamine.
Antimony glance	Stibnite.	Epsom salts	Magnesium sulphate.
Aqua fortis	Nitric acid.		
Aqua regia	Nitric and hydrochloric acids.	Feather ore	Jamesonite.
		Fibrolite	Sillimanite.
Argentine	Labradorite.	Fluorspar	Fluorite.
Arsenical nickel	Niccolite.	Fool's gold	Chalcopyrite, pyrite.
Asphalt	Asphaltum.		
Baryta	Barium oxide.	Fowler's solution	Potassium arsenite.
Barytes	Barium sulphate.		
		Galena	Galenite.
Beeswax mineral	Ozocerite.	Genthite	Garnierite.
Bell metal ore	Stannite.	Glauber's salts	Sodium sulphate.
Bitumen	Asphaltum.		
Black hematite	Psilomelane.	Gray antimony	Stibnite.
Black jack	Sphalerite.	Gray copper	Tetrahedrite, Tennantite.
Black lead	Graphite.		
Black mica	Biotite.	Green carbonate of copper.	Malachite.
Black oxide of manganese.	Pyrolusite.	Green vitriol	Iron sulphate.
		Gypsum	Calcium sulphate.
Blende	Sphalerite.		
Blue carbonate of copper.	Azurite.	Hartshorn	Ammonia water.
		Heavy spar	Barite.
Blue copper	Covellite and others.	Hematite black	Psilomelane.
		Hematite brown	Limonite.
Blue iron earth	Vivianite.	Horn silver	Cerargyrite, silver chloride.
Blue stone	Copper sulphate.		
Blue vitriol	Copper sulphate.	Horseflesh ore	Bornite, chalcopyrite.
Borax	Sodium borate.		
Brimstone	Sulphur.	Hypo	Sodium hyposulphite.
Brittle silver	Stephanite.		
Brown hemitite	Brucite.	Idocrase	Vesuvianite.
Butter of antimony	Antimonious chloride.	Indigo copper	Covellite.
		Iron pyrites	Pyrite, Marcasite.
Calamine electric	Calamine.		
Calc-spar	Calcite.	Isinglass	Muscovite.
Calomel	Mercurous chloride.	Jack	Sphalerite.
		Kaolin	Kaolinite.
Capillary pyrite	Millerite.	Kyanite	Cyanite.
Chalk	Calcium carbonate.	Laughing gas	Nitrous oxide.
		Leucopyrite	Lollingite.
China clay	Kaolinite.	Light ruby silver	Proustite.
Chrome iron ore	Chromite.	Lime	Calcium oxide.
Cobalt glance	Cobaltite.	Lime feldspar	Anorthite.
Common salt	Halite.	Lime water	Calcium hydrate.
Copalite	Amber.		
Copperas	Iron sulphate.	Litharge	Lead oxide.
Copper glance	Chalcocite.	Lithia mica	Lepidolite.
Copper nickel	Niccolite.	Liver of sulphur	Potassium sulphide.
Copper pyrite	Chalcopyrite.		
Cordierite	Iolite.	Lunar caustic	Silver nitrate.
Corrosive sublimate	Mercuric chloride.	Magnesia	Magnesium oxide.
Cream of tartar	Potassium bitartrate.	Magnesium limestone.	Dolomite.
Dark ruby silver	Pyrargyrite.		

TABLE 12.—*Common and technical names*—Continued

Common name	Technical name	Common name	Technical name
Magnesium mica	Biotite.	Ruby copper	Cuprite.
Magnetic iron ore	Magnetite.	Ruby jack	Sphalerite.
Magnetic pyrites	Pyrrhotite.	Ruby silver dark	Pyrargyrite.
Manganese blende	Alabandite.	Ruby silver light	Proustite.
Manganese glance	Alabandite.	Sal ammoniac	Ammonium chloride.
Marmatite	Ferriferous sphalerite.	Sal soda	Sodium carbonate, crystal.
Meerschaum	Magnesium silicate.	Sal volatile	Ammonium bicarbonate.
Menaccanite	Ilmenite.	Salt cake	Sodium sulphate.
Mercury blende	Cinnabar.		
Mica	Muscovite.	Salt, common	Sodium chloride.
Mine vermilion	Cinnabar.		
Mispickle	Arsenopyrite.	Saltpeter	Potassium nitrate.
Molly	Molybdenite.		
Mosaic gold	Tin bisulphide.	Scapolite	Wernerite.
Mundic	Pyrrhotite.	Schorl	Tourmaline.
Needle zeolite	Natrolite.	Silver glance	Argentite.
Nepheline	Nephelite.	Soda ash	Sodium carbonate.
Nickel arsenite	Niccolite.		
Nigrine	Rutile.	Soda feldspar	Albite.
Niter	Sodium nitrate.	Spathic ore	Siderite.
Niter cake	Sodium bisulphate.	Specular iron	Hematite.
		Spirits of salt	Hydrochloric acid.
Noumeite	Garnierite.		
Oil of vitriol	Sulphuric acid.	Sugar of lead	Lead acetate.
Olivine	Chrysolite.	Tartar emetic	Antimony and potassium tartrate.
Peacock ore	Bornite, chalcopyrite.		
Pencil stone	Pyrophyllite.	Tinkal	Borax.
Pericline	Albite.	Tin pyrite	Stannite.
Pitch blende	Uraninite.	Titanic iron ore	Ilmenite.
Plumbago	Graphite.	Tungsten	Wolframite, huebnerite.
Potash feldspar	Orthoclase, microline.		
Potash mica	Muscovite.	Variegated copper	Bornite.
Prussian blue	Ferric-ferrocyanide.	Verdigris	Copper acetate.
		Vermilion	Cinnabar.
Prussic acid	Hydrocyanic acid.	Vinegar	Acetic acid.
		Water glass	Sodium silicate.
Purple copper ore	Bornite.	White iron	Pyrite, marcasite.
Pyro	Pyrogallic acid.		
Quicksilver	Mercury.	White lead	Lead carbonate.
Red iron ore	Hematite.	White lead ore	Cerussite
Red lead	Lead oxide.	White mica	Muscovite.
Red ochre	Hematite.	White precipitate	Ammonium mercuric chloride.
Red oxide of copper	Cuprite.		
Red precipitate	Red mercuric oxide.	White vitriol	Zinc sulphate.
Red silver	Proustite.	White zinc	Zinc oxide.
Red zinc ore	Zincite.	Yellow jack	Sphalerite.
Ripidolite	Chlorite.	Zinc	Sphalerite.
Rochelle salts	Potassium and sodium tartrate.	Zinc blende	Sphalerite.
		Zinc bloom	Hydrozincite.
		Zinc spinel	Gahnite.
Rock salt	Halite.		

94

TABLE 13.—*Approximate percentages of metals recovered*

Metal	Concentration		Amalga- mation	Cyanida- tion	Smelting
	Gravity	Flotation			
Gold	80–90	80–90	70–95	70–95	100
Silver	60–80	80–95	60–80	60–90	95–98
Copper	60–80	90–95			90–95
Lead	60–80	80–95			90–95
Zinc	60–80	80–95			80–90
Iron	60–75				95
Tin	60–80				90–95

TABLE 14.—*Counting colors*

For estimating purposes, color sizes are assumed to weigh:
No. 1 colors—3.0 mg. or over.
No. 2 colors—1.33 mg. to 3.0 mg.
No. 3 colors—0.333 mg. to 1.33 mg.
No. 4 colors—0.02 mg. to 0.333 mg.   (Includes finest colors.)
In estimation, weight of colors is taken as the first figure given above, except for No. 4 colors, which are counted as 50 per mg. An expert panner should come within 5% of the final balance weight after the gold has been acid-cleaned.

170 colors/1¢ = 314,500 colors/troy oz.
280 colors/1¢ = 436,900 colors/troy oz.
500 colors/1¢ = 885,000 colors/troy oz.

Coins	Weight— new (grams)	Weight— very worn (grams)
Copper cent	3.1103	2.8
Nickel	4.999	4.5
Dime	2.5000	2.3
Quarter	6.2500	5.5
Half dollar	12.500	11.7
Silver dollar	26.730	25.0

TABLE 15.—*Weights and measures*

Lengths

1 mile = $\begin{cases} \text{8 furlongs} \\ \text{80 chains} \\ \text{320 rods} \\ \text{1,760 yards} \\ \text{5,280 feet} \end{cases}$

1 furlong = $\begin{cases} \text{10 chains} \\ \text{220 yards} \end{cases}$

1 station = $\begin{cases} \text{6.06 rods} \\ \text{33.3 yards} \\ \text{100 feet} \end{cases}$

1 chain = $\begin{cases} \text{4 rods} \\ \text{22 yards} \\ \text{66 feet} \\ \text{100 links} \end{cases}$

1 rod = $\begin{cases} \text{5.5 yards} \\ \text{16.5 feet} \end{cases}$

1 yard = $\begin{cases} \text{3 feet} \\ \text{36 inches} \end{cases}$

1 vara = 33 inches (approx.)
1 foot = 12 inches
1 link = 7.92 inches
1 inch = 0.0833 foot

Square Measure

1 township = 36 sq. miles

1 sq. mile = $\begin{cases} \text{1 section} \\ \text{640 acres} \end{cases}$

1 acre = $\begin{cases} \text{4,840 sq. yards} \\ \text{43,560 sq. feet} \\ \text{10 sq. chains} \\ \text{160 sq. rods} \end{cases}$

Lode Claim = $\begin{cases} \text{600 ft. x 1,500 ft.} \\ \text{20.661 acres} \\ \text{3305.78 sq. rods} \end{cases}$

Placer Claim = 20 acres
(1 locator)
1 sq. rod = 272¼ sq. feet
1 sq. yard = 9 sq. feet
1 sq. foot = 144 sq. inches

Cubic Measure

1 cubic yard = 27 cubic feet
1 cord (wood) = 4 x 4 x 8 ft. = 128 cu. ft.
1 ton (shipping) = 40 cubic ft.
1 cubic foot = 1,728 cubic inches

1 cu. ft. = 7.48 U.S. gallons
1 bushel = 2150.42 cu. in.
1 bushel = 1.244 + cu. ft.
1 gallon = 231 cu. in.

Weights (Commercial)

1 long ton = 2,240 lbs.
1 short ton = 2,000 lbs.

1 pound = 16 ounces
1 ounce = 16 drams

Troy Weight (for Gold and Silver)

12 oz. troy = 1 lb. troy = 0.823 pounds av. = 5760 grains
16 oz. av. = 1 lb. av. = 7,000 grains
31.1035 grams = 1 oz. troy = 20 pennyweight = 480 grains
28.350 grams = 1 oz. av.
1 kilogram = 2.2046 lb. av.
1 pennyweight = 24 grains = 1.555 grams

Dry Measure

1 bushel = $\begin{cases} \text{4 pecks} \\ \text{32 quarts} \end{cases}$

1 peck = 8 quarts

1 quart = 2 pints
1 bushel = 1.2445 cubic feet

Liquid Measure

1 barrel=31½ gallons
1 gallon=4 quarts
1 quart=2 pints

1 pint= $\begin{cases} 4 \text{ gills} \\ 2 \text{ cups} \end{cases}$
1 gallon=231 cubic inches

Miscellaneous

1.8° Fahrenheit=1° Centigrade
14.696 pounds per square inch=atmospheric pressure, sea level
13.144 cubic feet=volume 1 lb. air at 62° F. and 14.7 pounds per sq. inch
1 gallon water=8⅓ lbs. at 62° F.
1 cu. ft. water=62.5 lbs. at 62° F.
2.304 depth of water=1 lb. per sq. in.
Doubling the diameter of pipe or hose increases its capacity approximately four times
Diameter of circle: multiply circumference by .31831
Circumference of circle: multiply diameter by 3.1416
Area of circle: multiply square of diameter by .7854—multiply square of radius by 3.1416

TABLE 16.—*Weight of rocks*

Material	Wt. per cu. ft., lb.		Cu. ft. per ton		Tons per cu. yd.	
	In place	Broken	In place	Broken	In place	Broken
Dolomite	160		12.5		2.16	1.30
Gneiss	168	96	11.9	20.8	2.27	1.30
Granite and porphyry	170	97	11.8	20.6	2.30	1.31
Greenstone and trap	187	107	10.7	18.7	2.52	1.39
Hematite*	267		7.5		3.60	
Limestone	168	96	11.9	20.8	2.27	1.30
Limestone ores*	154		13.0		2.08	
Quartz	165	94	12.1	21.3	2.23	1.27
Quartzose ores*	138		14.5		1.86	
Sandstone	151	86	13.2	23.3	2.08	1.16
Slate	175	95	11.4	21.1	2.36	1.28
Vein quartz*	148		13.5		2.00	
Vein quartz, 15% PbS*	164		12.2		2.21	
Vein quartz, 15% FeS₂*	160		12.5		2.16	

*Refers to possible wt. of ores; pure minerals usually weigh more.

Swelling in fill.—On excavating a mixture of solid and loose rock and earth, 1 cu. yd. in place makes about 1.4 cu. yd. in fill. If rock be first stripped of earth, and then blasted and dumped by itself, the percentage of voids is larger. At Boulder, Colo., 3,600 cu. yd. of solid rock made a 5,340 cu. yd. embankment; a ratio of 1:1.51. In Virginia, 50,000 cu. yd. of limestone and mica schist, broken and put in embankment, made 90,000 cu. yd. an increase of 80%. In subaqueous excavation.—Ashtabula Harbor, 62,869 cu. yd. (place measure) gave 103,537 cu. yd. measured in scows, an increase of 65%. 18 cubic feet of earth or gravel in place equals approximately 27 cubic feet when excavated.

TABLE **17.**—*Conversion tables*

Metric and U.S. Weights and Measures

Lengths

Miles×1.6093=Kilometers
Yards×.9144=Meters
Feet×.3048=Meters
Feet×30.48=Centimeters
Inches×2.54=Centimeters
Inches×25.4=Millimeters
Kilometers×.621=Miles

Kilometers×1093.6=Yards
Kilometers×3280.9=Feet
Meters×1.094=Yards
Meters×3.281=Feet
Meters×39.37=Inches
Centimeters×.3937=Inches
Millimeters×.03937=Inches

Areas

Sq. mile×2.59=Sq. kilometers
Acres×.00405=Sq. kilometers
Acres×.4047=Hectares
Sq. Yards×.8361=Sq. meters
Sq. feet×.0929=Sq. meters
Sq. Inches×6.452=Sq. centimeters
Sq. inches×645.2=Sq. millimeters
Sq. kilometers×.3861=Sq. miles
Sq. kilometers×247.11=Acres
Hectares×2.471=Acres
Sq. meters×1.196=Sq. yards
Sq. meters×10.764=Sq. feet
Sq. centimeters×.155=Sq. inches
Sq. millimeters×.00155=Sq. inches

Volume

Cu. yards×.765=Cu. meters
Cu. feet×.0283=Cu. meters
Cu. inches×16.383=Cu. centimeters
Cu. meters×1.308=Cu. yards
Cu. meters×35.3145=Cu. feet
Cu. centimeters×.06102=Cu. inches

Liquid Measure

U.S. gallons×.8333=Imperial gallons
Gallons×3.785=Litres
Quarts×.946=Litres
Imperial gallons×1.2009=U.S. gallons
Litres×.2642=Gallons
Litres×1.057=Quarts

Weights

Pounds×.453=Kilograms

Kilograms×2.2046=Pounds

TABLE 18.—*Water equivalent*

1 gallon (gal.) = 231 cubic inches
1 gallon (gal.) = .1337 cubic feet
1 gallon of water weighs 8.33 pounds
1 million gallons (m.g.) = 3.0689 acre-feet
1 cubic foot (cu. ft.) = 1728 cubic inches
1 cubic foot (cu. ft.) = 7.48 gallons
1 cubic foot of water weighs 62.4 pounds
1 acre foot (ac. ft.) = amount of water required to cover one acre one foot deep
1 acre foot (ac. ft.) = 43,560 cubic feet
1 acre foot (ac. ft.) = 325,850 gallons
1 acre foot (ac. ft.) = 12 acre inches
1 gallon per minute (g.p.m.) = 0.00223 cubic feet per second
1 gallon per minute (g.p.m.) = 1,440 gallons per day (24 hours)
1 million gallons per 24 hours (m.g.d.) = 1.547 cubic feet per second
1 million gallons per 24 hours (m.g.d.) = 695 gallons per minute
1 cubic foot per second (c.f.s.) = 7.48 gallons per second
1 cubic foot per second (c.f.s.) = 448.83 gallons per minute
1 cubic foot per second (c.f.s.) = 646,272 gallons per day (24 hours)
1 cubic foot per second (c.f.s.) = .992 acre inch per hour
1 cubic foot per second (c.f.s.) = 1.983 acre feet per day (24 hours)

TABLE 19.—*Conversion of units of flow*

Cubic feet per second	Gallons per minute	Million gallons per day	Miner's inches			Acre-inches per hour	Acre-feet per 24 hours
			Arizona, California, Montana, Oregon	Idaho, Kansas, Nebraska, Nevada, New Mexico, North Dakota, South Dakota, Utah	Colorado		
1	448.8	0.646	40	50	38.4	0.992	1.983
0.00223	1	0.001440	0.0891	0.1114	0.0856	0.0022	0.00442
1.547	694.4	1	61.89	77.36	59.44	1.535	3.07
0.025	11.25	0.0162	1	1.25	0.960	0.0248	0.0496
0.020	9.00	0.01296	0.80	1	0.768	0.0198	0.0397
0.026	11.69	0.0168	1.042	1.302	1	0.0258	0.0516
1.01	452.42	0.651	40.32	50.40	38.71	1	2.00
.504	226.3	0.3258	20.17	25.21	19.36	0.5	1

99

TABLE 20.—*Mill water requirements*

Tons ore fed per 24 hours	Dilutions							
	3 : 1		4 : 1		5 : 1		6 : 1	
	Tons water per 24 hrs.	Gals. water per min.	Tons water per 24 hrs.	Gals. water per min.	Tons water per 24 hrs.	Gals. water per min.	Tons water per 24 hrs.	Gals. water per min.
10	30	5.0	40	6.7	50	8.3	60	10.0
15	45	7.5	60	10.0	75	12.5	90	15.0
20	60	10.0	80	13.3	100	16.6	120	20.0
25	75	12.5	100	16.6	125	20.8	150	25.0
35	105	17.5	140	23.3	175	29.3	210	35.0
50	150	25.0	200	33.3	250	41.7	300	50.0
65	195	32.5	260	43.3	325	54.2	390	65.0
100	300	50.0	400	66.7	500	83.4	600	100.0
125	375	62.5	500	83.4	625	104.0	750	125.0
150	450	75.0	600	100.0	750	125.0	900	150.0
200	600	100.0	800	133.3	1,000	166.6	1,200	200.0
300	900	150.0	1,200	200.0	1,500	250.0	1,800	300.0
500	1,500	250.0	2,000	333.3	2,500	417.0	3,000	500.0

Machines	Water to ore ratios	Machines	Water to ore ratios
Harz type jigs	3 : 1 to 5 : 1	Filter discharge	1 : 20 to 1 : 5
Flotation machines (oxidized ore)	2 : 1 to 3 : 1	Stamps—Gravity	3 : 1 to 6 : 1
		Ball and rod mills	1 : 1 to 1 : 4
Flotation machines (sulphide ore)	3 : 1 to 5 : 1	Hydraulic classifiers	3 : 1 to 10 : 1
		Screens	1 : 1 to 3 : 1
Cyanide slime agitators	1 : 1 to 2 : 1	Concentrating tables	2 : 1 to 5 : 1
Thickener discharge	1 : 1 to 2 : 1	Denver mineral jigs	1 : 1 to 2 : 1

Average cyanide circuits 1 to 3 tons water : ton ore.
Average flotation circuits 3 to 5 tons water : ton ore.
Average table circuits 5 to 7 tons water : ton ore.
Average jig and table circuits 6 to 10 tons water : ton ore.
Average table and amalgam. 8 to 12 tons water : ton ore.
1 gallon water=8.33 pounds=3.785 liters.
1 ton water=240 gallons=908.49 liters.
1 cubic foot=7.48 gallons.
G.P.M.=Tons of water per 24 hours×0.16643.

TABLE 21.—*Chains and links to feet*

Links	0 chs.	1 ch.	2 chs.	3 chs.	4 chs.	Links	0 chs.	1 ch.	2 chs.	3 chs.	4 chs.
	Feet	Feet	Feet	Feet	Feet		Feet	Feet	Feet	Feet	Feet
0	0.00	66.00	132.00	198.00	264.00	50	33.00	99.00	165.00	231.00	297.00
1	0.66	66.66	132.66	198.66	264.66	51	33.66	99.66	165.66	231.66	297.66
2	1.32	67.32	133.32	199.32	265.32	52	34.32	100.32	166.32	232.32	298.32
3	1.98	67.98	133.98	199.98	265.98	53	34.98	100.98	166.98	232.98	298.98
4	2.64	68.64	134.64	200.64	266.64	54	35.64	101.65	167.64	233.64	299.64
5	3.30	69.30	135.30	201.30	267.30	55	36.30	102.30	168.30	234.30	300.30
6	3.96	69.96	135.96	201.96	267.96	56	36.96	102.96	168.96	234.96	300.96
7	4.62	70.62	136.62	202.62	268.62	57	37.62	103.62	169.62	235.62	301.62
8	5.28	71.28	137.28	203.28	269.28	58	38.28	104.28	170.28	236.28	302.28
9	5.94	71.94	137.94	203.94	269.94	59	38.94	104.94	170.94	236.94	302.94
10	6.60	72.60	138.60	204.60	270.60	60	39.60	105.60	171.60	237.60	303.60
11	7.26	73.26	139.26	205.26	271.26	61	40.26	106.26	172.26	238.26	304.26
12	7.92	73.92	139.92	205.92	271.92	62	40.92	106.92	172.92	238.92	304.92
13	8.58	74.58	140.58	206.58	272.58	63	41.58	107.58	173.58	239.58	305.58
14	9.24	75.24	141.24	207.24	273.24	64	42.24	108.24	174.24	240.24	306.24
15	9.90	75.90	141.90	207.90	273.90	65	42.90	108.90	174.90	240.90	306.90
16	10.56	76.56	142.56	208.56	274.56	66	43.56	109.56	175.56	241.56	307.56
17	11.22	77.22	143.22	209.22	275.22	67	44.22	110.22	176.22	212.22	308.22
18	11.88	77.88	143.88	209.88	275.88	68	44.88	110.88	176.88	242.88	308.88
19	12.54	78.54	144.54	210.54	276.54	69	45.54	111.54	177.54	213.54	309.54
20	13.20	79.20	145.20	211.20	277.20	70	46.20	112.20	178.20	244.20	310.20
21	13.86	79.86	145.86	211.86	277.86	71	46.86	112.86	178.86	244.86	310.86
22	14.52	80.52	146.52	212.52	278.52	72	47.52	113.52	179.52	245.52	311.52
23	15.18	81.18	147.18	213.18	279.18	73	48.18	114.18	180.18	246.18	312.18
24	15.84	81.84	147.84	213.84	279.84	74	48.84	114.84	180.84	246.84	312.84
25	16.50	82.50	148.50	214.50	280.50	75	49.50	115.50	181.50	247.50	313.50
26	17.16	83.16	149.16	215.16	281.16	76	50.16	116.16	182.16	248.16	314.16
27	17.82	83.82	149.82	215.82	281.82	77	50.82	116.82	182.82	248.82	314.82
28	18.48	84.48	150.48	216.48	282.48	78	51.48	117.48	183.48	249.48	315.48
29	19.14	85.14	151.14	217.14	283.14	79	52.14	118.14	184.14	250.14	316.14
30	19.80	85.80	151.80	217.80	283.80	80	52.80	118.80	184.80	250.80	316.80
31	20.46	86.46	152.46	218.46	284.46	81	53.46	119.46	185.46	251.46	317.46
32	21.12	87.12	153.12	219.12	285.12	82	54.12	120.12	186.12	252.12	318.12
33	21.78	87.78	153.78	219.78	285.78	83	54.78	120.78	186.78	252.78	318.78
34	22.44	88.44	154.44	220.44	286.44	84	55.44	121.44	187.44	253.44	319.44
35	23.10	89.10	155.10	221.10	287.10	85	56.10	122.10	188.10	254.10	320.10
36	23.76	89.76	155.76	221.76	287.76	86	56.76	122.76	188.76	254.76	320.76
37	24.42	90.42	156.42	222.42	288.42	87	57.42	123.42	189.42	255.42	321.42
38	25.08	91.08	157.08	223.08	289.08	88	58.08	124.08	190.08	256.08	322.08
39	25.74	91.74	157.74	223.74	289.74	89	58.74	124.74	190.74	256.74	322.74
40	26.40	92.40	158.40	224.40	290.40	90	59.40	125.40	191.40	257.40	323.40
41	27.06	93.06	159.06	225.06	291.06	91	60.06	126.06	192.06	258.06	324.06
42	27.72	93.72	159.72	225.72	291.72	92	60.72	126.72	192.72	258.72	324.72
43	28.38	94.38	160.38	226.38	292.38	93	61.38	127.38	193.38	259.38	325.38
44	29.04	95.04	161.04	227.04	293.04	94	62.04	128.04	194.04	260.04	326.04
45	29.70	95.70	161.70	227.70	293.70	95	62.70	128.70	194.70	260.70	326.70
46	30.36	96.36	162.36	228.36	294.36	96	63.36	129.36	195.36	261.36	327.36
47	31.02	97.02	163.02	229.02	295.02	97	64.02	130.02	196.02	262.02	328.02
48	31.68	97.68	163.68	229.68	295.68	98	64.68	130.68	196.68	262.68	328.68
49	32.34	98.34	164.34	230.34	296.34	99	65.34	131.34	197.34	263.34	329.34
50	33.00	99.00	165.00	231.00	297.00	100	66.00	132.00	198.00	264.00	330.00

For conversion of even or whole chains, up to 100 chains, use columns zero chains and links, and place decimal two points to the right.

TABLE 21.—*Chains and links to feet—Continued*

Links	5 chs.	6 chs.	7 chs.	8 chs.	9 chs.	Links	5 chs.	6 chs.	7 chs.	8 chs.	9 chs.
	Feet	*Feet*	*Feet*	*Feet*	*Feet*		*Feet*	*Feet*	*Feet*	*Feet*	*Feet*
0	330.00	396.00	462.00	528.00	594.00	50	363.00	429.00	495.00	561.00	627.00
1	330.66	396.66	462.66	528.66	594.66	51	363.66	429.66	495.66	561.66	627.66
2	331.32	397.32	463.32	529.32	595.32	52	364.32	430.32	496.32	562.32	628.32
3	331.98	397.98	463.98	529.98	595.98	53	364.98	430.98	496.98	562.98	628.98
4	332.64	398.64	464.64	530.64	596.64	54	365.64	431.64	497.64	563.64	629.64
5	333.30	399.30	465.30	531.30	597.30	55	366.30	432.30	498.30	564.30	630.30
6	333.96	399.96	465.96	531.96	597.96	56	366.96	432.96	498.96	564.96	630.96
7	334.62	400.62	466.62	532.62	598.62	57	367.62	433.62	499.62	565.62	631.62
8	335.28	401.28	467.28	533.28	599.28	58	368.28	434.28	500.28	566.28	632.28
9	335.94	401.94	467.94	533.94	599.94	59	368.94	434.94	500.94	566.94	632.94
10	336.60	402.60	468.60	534.60	600.60	60	369.60	435.60	501.60	567.60	633.60
11	337.26	403.26	469.26	535.26	601.26	61	370.26	436.26	502.26	568.26	634.26
12	337.92	403.92	469.92	535.92	601.92	62	370.92	436.92	502.92	568.92	634.92
13	338.58	404.58	470.58	536.58	602.58	63	371.58	437.58	503.58	569.58	635.58
14	339.24	405.24	471.24	537.24	603.24	64	372.24	438.24	504.24	570.24	636.24
15	339.90	405.90	471.90	537.90	603.90	65	372.90	438.90	504.90	570.90	636.90
16	340.56	406.56	472.56	538.56	604.56	66	373.56	439.56	505.56	571.56	637.56
17	341.22	407.22	473.22	539.22	605.22	67	374.22	440.22	506.22	572.22	638.22
18	341.88	407.88	473.88	539.88	605.88	68	374.88	440.88	506.88	572.88	638.88
19	342.54	408.54	474.54	540.54	606.54	69	375.54	441.54	507.54	573.54	639.54
20	343.20	409.20	475.20	541.20	607.20	70	376.20	442.20	508.20	574.20	640.20
21	343.86	409.86	475.86	541.86	607.86	71	376.86	442.86	508.86	574.86	640.86
22	344.52	410.52	476.52	542.52	608.52	72	377.52	443.52	509.52	575.52	641.52
23	345.18	411.18	477.18	543.18	609.18	73	378.18	444.18	510.18	576.18	642.18
24	345.84	411.84	477.84	543.84	609.84	74	378.84	444.84	510.84	576.84	642.84
25	346.50	412.50	478.50	544.50	610.50	75	379.50	445.50	511.50	577.50	643.50
26	347.16	413.16	479.16	545.16	611.16	76	380.16	446.16	512.16	578.16	644.16
27	347.82	413.82	479.82	545.82	611.82	77	380.82	446.82	512.82	578.82	644.82
28	348.48	414.48	480.48	546.48	612.48	78	381.48	447.48	513.48	579.48	645.48
29	349.14	415.14	481.14	547.14	613.14	79	382.14	448.14	514.14	580.14	646.14
30	349.80	415.80	481.80	547.80	613.80	80	382.80	448.80	514.80	580.80	646.80
31	350.46	416.46	482.46	548.46	614.46	81	383.46	449.46	515.46	581.46	647.46
32	351.12	417.12	483.12	549.12	615.12	82	384.12	450.12	516.12	582.12	648.12
33	351.78	417.78	483.78	549.78	615.78	83	384.78	450.78	516.78	582.78	648.78
34	352.44	418.44	484.44	550.44	616.44	84	385.44	451.44	517.44	583.44	649.44
35	353.10	419.10	485.10	551.10	617.10	85	386.10	452.10	518.10	584.10	650.10
36	353.76	419.76	485.76	551.76	617.76	86	386.76	452.76	518.76	584.76	650.76
37	354.42	420.42	486.42	552.42	618.42	87	387.42	453.42	519.42	585.42	651.42
38	355.08	421.08	487.08	553.08	619.08	88	388.08	454.08	520.08	586.08	652.08
39	355.74	421.74	487.74	553.74	619.74	89	388.74	454.74	520.74	586.74	652.74
40	356.40	422.40	488.40	554.40	620.40	90	389.40	455.40	521.40	587.40	653.40
41	357.06	423.06	489.06	555.06	621.06	91	390.06	456.06	522.06	588.06	654.06
42	357.72	423.72	489.72	555.72	621.72	92	390.72	456.72	522.72	588.72	654.72
43	358.38	424.38	490.38	556.38	622.38	93	391.38	457.38	523.38	589.38	655.38
44	359.04	425.04	491.04	557.04	623.04	94	392.04	458.04	524.04	590.04	656.04
45	359.70	425.70	491.70	557.70	623.70	95	392.70	458.70	524.70	590.70	656.70
46	360.36	426.36	492.36	558.36	624.36	96	393.36	459.36	525.36	591.36	657.36
47	361.02	427.02	493.02	559.02	625.02	97	394.02	460.02	526.02	592.02	658.02
48	361.68	427.68	493.68	559.68	625.68	98	394.68	460.68	526.68	592.68	658.68
49	362.34	428.34	494.34	560.34	626.34	99	395.34	461.34	527.34	593.34	659.34
50	363.00	429.00	495.00	561.00	627.00	100	396.00	462.00	528.00	594.00	660.00

For conversion of even or whole chains, up to 1000 chains, use columns chains and links, and place decimal two points to the right.

TABLE 22.—*Direct mine operating costs for various mining methods. Period—1955–1959*

| Mining method | Tons mined per month [1] | Direct mining costs/ton [2] | | | Labor—percent of total cost [3] |
		High	Low	Average [3]	
Squaresetting	439,330	$18.72	$6.22	$10.20	71.2
Cut and fill	585,300	14.73	3.07	6.69	56.7
Shrinkage	305,820	8.12	1.75	3.92	N.A.
Room and pillar (trackless type)	733,220	2.41	1.16	2.05	43.7
Sub-level stoping	1,547,410	4.71	1.06	2.37	56.9
Sub-level caving	118,150	N.A.	N.A.	4.97	63.3
Block caving	1,803,150	2.25	1.15	[5] 1.41	54.2
Open pit [4]	5,198,060	1.15	.21	.32	36.4

[1] Total aggregate tonnage of ore mined each month except for open pits (see footnote 4).
[2] Direct mining costs include (a) exploration and development, (b) stoping, (c) haulage, (d) hoisting, (e) pumping, and (f) general underground and surface.
[3] Weighted average on the basis of tons produced from each mine.
[4] Cost is per ton of "material" and is based on total tons of ore and waste handled.
[5] This average may be on the high side due to lack of information covering a number of efficient operations.
N.A.—Not available.

TABLE 23.—*Schedule of prices for uranium ore*

Grade of ore, percent U₃O₈	Pounds of U₃O₈ per ton of ore	Base price		Grade premium		Mine development allowance .50/lb.	Price per ton of ore		
		Pound of U₃O₈	Ton of ore	75¢ a lb. over 4-lb.	25¢ a lb. over 10-lb.		Price before initial production bonus and haulage allowance	Initial production bonus on 10,000 lbs.	Price before haulage allowance
0.10	2.00	$1.50	$3.00	--------	--------	$1.00	$4.00	$3.00	$7.00
.11	2.20	1.70	3.74	--------	--------	1.10	4.84	3.74	8.58
.12	2.40	1.90	4.56	--------	--------	1.20	5.76	4.56	10.32
.13	2.60	2.10	5.46	--------	--------	1.30	6.76	5.46	12.22
.14	2.80	2.30	6.44	--------	--------	1.40	7.84	6.44	14.28
.15	3.00	2.50	7.50	--------	--------	1.50	9.00	7.50	16.50
.16	3.20	2.70	8.64	--------	--------	1.60	10.24	8.64	18.88
.17	3.40	2.90	9.86	--------	--------	1.70	11.56	9.86	21.42
.18	3.60	3.10	11.16	--------	--------	1.80	12.96	11.16	24.12
.19	3.80	3.30	12.54	--------		1.90	14.44	12.54	26.98
.20	4.00	3.50	14.00			2.00	16.00	14.00	30.00
.21	4.20	3.50	14.70	$0.15	--------	2.10	16.95	14.70	31.65
.22	4.40	3.50	15.40	.30	--------	2.20	17.90	15.40	33.30
.23	4.60	3.50	16.10	.45	--------	2.30	18.85	16.10	34.95
.24	4.80	3.50	16.80	.60	--------	2.40	19.80	16.80	36.60
.25	5.00	3.50	17.50	.75	--------	2.50	20.75	17.50	38.25
.26	5.20	3.50	18.20	.90	--------	2.60	21.70	18.20	39.90
.27	5.40	3.50	18.90	1.05	--------	2.70	22.65	18.90	41.55
.28	5.60	3.50	19.60	1.20	--------	2.80	23.60	19.60	43.20
.29	5.80	3.50	20.30	1.35	--------	2.90	24.55	20.30	44.85
.30	6.00	3.50	21.00	1.50	--------	3.00	25.50	21.00	46.50
.31	6.20	3.50	21.70	1.65	--------	3.10	26.45	21.70	48.15
.32	6.40	3.50	22.40	1.80	--------	3.20	27.40	22.40	49.80
.33	6.60	3.50	23.10	1.95	--------	3.30	28.35	23.10	51.45
.34	6.80	3.50	23.80	2.10	--------	3.40	29.30	23.80	53.10
.35	7.00	3.50	24.50	2.25	--------	3.50	30.25	24.50	54.75
.36	7.20	3.50	25.20	2.40	--------	3.60	31.20	25.20	56.40
.37	7.40	3.50	25.90	2.55	--------	3.70	32.15	25.90	58.05
.38	7.60	3.50	26.60	2.70	--------	3.80	33.10	26.60	59.70
.39	7.80	3.50	27.30	2.85	--------	3.90	34.05	27.30	61.35
.40	8.00	3.50	28.00	3.00	--------	4.00	35.00	28.00	63.00
.41	8.20	3.50	28.70	3.15	--------	4.10	35.95	28.70	64.65
.42	8.40	3.50	29.40	3.30	--------	4.20	36.90	29.40	66.30
.43	8.60	3.50	30.10	3.45	--------	4.30	37.85	30.10	67.95
.44	8.80	3.50	30.80	3.60	--------	4.40	38.80	30.80	69.60
.45	9.00	3.50	31.50	3.75	--------	4.50	39.75	31.50	71.25
.46	9.20	3.50	32.20	3.90	--------	4.60	40.70	32.20	72.90
.47	9.40	3.50	32.90	4.05	--------	4.70	41.65	32.90	74.55
.48	9.60	3.50	33.60	4.20	--------	4.80	42.60	33.60	76.20
.49	9.80	3.50	34.30	4.35	--------	4.90	43.55	34.30	77.85
.50	10.00	3.50	35.00	4.50	--------	5.00	44.50	35.00	79.50
.60	12.00	3.50	42.00	6.00	$0.50	6.00	54.50	42.00	96.50
.70	14.00	3.50	49.00	7.50	1.00	7.00	64.50	49.00	113.50
.80	16.00	3.50	56.00	9.00	1.50	8.00	74.50	56.00	130.50
.90	18.00	3.50	63.00	10.50	2.00	9.00	84.50	63.00	147.50
1.00	20.00	3.50	70.00	12.00	2.50	10.00	94.50	70.00	164.50
2.00	40.00	3.50	140.00	27.00	7.50	20.00	194.50	140.00	334.50
3.00	60.00	3.50	210.00	42.00	12.50	30.00	294.50	210.00	504.50
4.00	80.00	3.50	280.00	57.00	17.50	40.00	394.50	280.00	674.50
5.00	100.00	3.50	350.00	72.00	22.50	50.00	494.50	350.00	844.50
6.00	120.00	3.50	420.00	87.00	27.50	60.00	594.50	420.00	1,014.50
7.00	140.00	3.50	490.00	102.00	32.50	70.00	694.50	490.00	1,184.50
8.00	160.00	3.50	560.00	117.00	37.50	80.00	794.50	560.00	1,354.50
9.00	180.00	3.50	630.00	132.00	42.50	90.00	894.50	630.00	1,524.50
10.00	200.00	3.50	700.00	147.00	47.50	100.00	994.50	700.00	1,694.50

104

Selected List of References

The following list is not all-inclusive, but contains many of the well known references on mining and related subjects that are generally available. No attempt has been made to group the references by subject matter, the titles being self-descriptive. For ready reference, the list is arranged alphabetically by author.

Bateman, A. M., *Economic Mineral Deposits*, John Wiley & Sons, Inc., New York, 2d ed., 1950.

Billings, M. P., *Structural Geology*, Prentice-Hall, New York, 1942.

Bureau of Mines, *Mineral Facts and Problems*, Government Printing Office, Bulletin 585, 1960.

Dana, E. S., and Ford, W. E., *A Textbook of Mineralogy*, John Wiley & Sons, Inc., New York, 4th ed., 1932.

Fay, A. H., *A Glossary of the Mining and Mineral Industry*, Bureau of Mines Bulletin 95, Government Printing Office, 1920.

Forrester, J. D., *Principles of Field and Mining Geology*, John Wiley & Sons, Inc., New York, 1955.

Hoover, T. J., *The Economics of Mining*, Stanford University Press, Stanford, California, 3d ed., 1948.

Jackson, C. F. and Knaebel, J. B., *Sampling and Estimation of Ore Deposits*, Bureau of Mines Bulletin 365, 1934.

Ladoo, R. B. and Myers, W. M., *Nonmetallic Minerals*, McGraw-Hill Book Co., Inc., New York, 2d ed., 1951.

Lahee, F. H., *Field Geology*, McGraw-Hill Book Co., Inc., New York, 4th ed., 1941.

Lewis, R. S., *Elements of Mining*, John Wiley & Sons, Inc., New York, 2d ed., 1958.

Lindgren, W., *Mineral Deposits*, McGraw-Hill Book Co., Inc., 4th ed., 1933.

Lindley, C. H., *Lindley on Mines*, Bancroft-Whitney Co., San Francisco, 3d ed., 1914.

Marston, A., Winfrey, R. and Hempstead, J. C., *Engineering Valuation and Depreciation*, McGraw-Hill Book Co., Inc., New York, 2d ed., 1953.

McKinstry, H. E., *Mining Geology*, Prentice-Hall, New York, 1948.

Parks, R. D., *Examination and Valuation of Mineral Property*, Addison-Wesley Publishing Co., Inc., Reading, Mass., 4th ed., 1957.

Ricketts, A. H., *American Mining Law*, California Division of Mines Bulletin 123, San Francisco, 4th ed., 1943.

Taggart, A. F., *Handbook of Mineral Dressing*, John Wiley & Sons, Inc., New York, 1945.

Von Bernewitz, M. W., *Handbook for Prospectors and Operators of Small Mines*, McGraw-Hill Book Co., Inc., New York, 4th ed., 1943.

Young, G. J., *Elements of Mining*, McGraw-Hill Book Co., Inc., New York, 4th ed., 1943.

* U.S. GOVERNMENT PRINTING OFFICE : 1962 O—642918

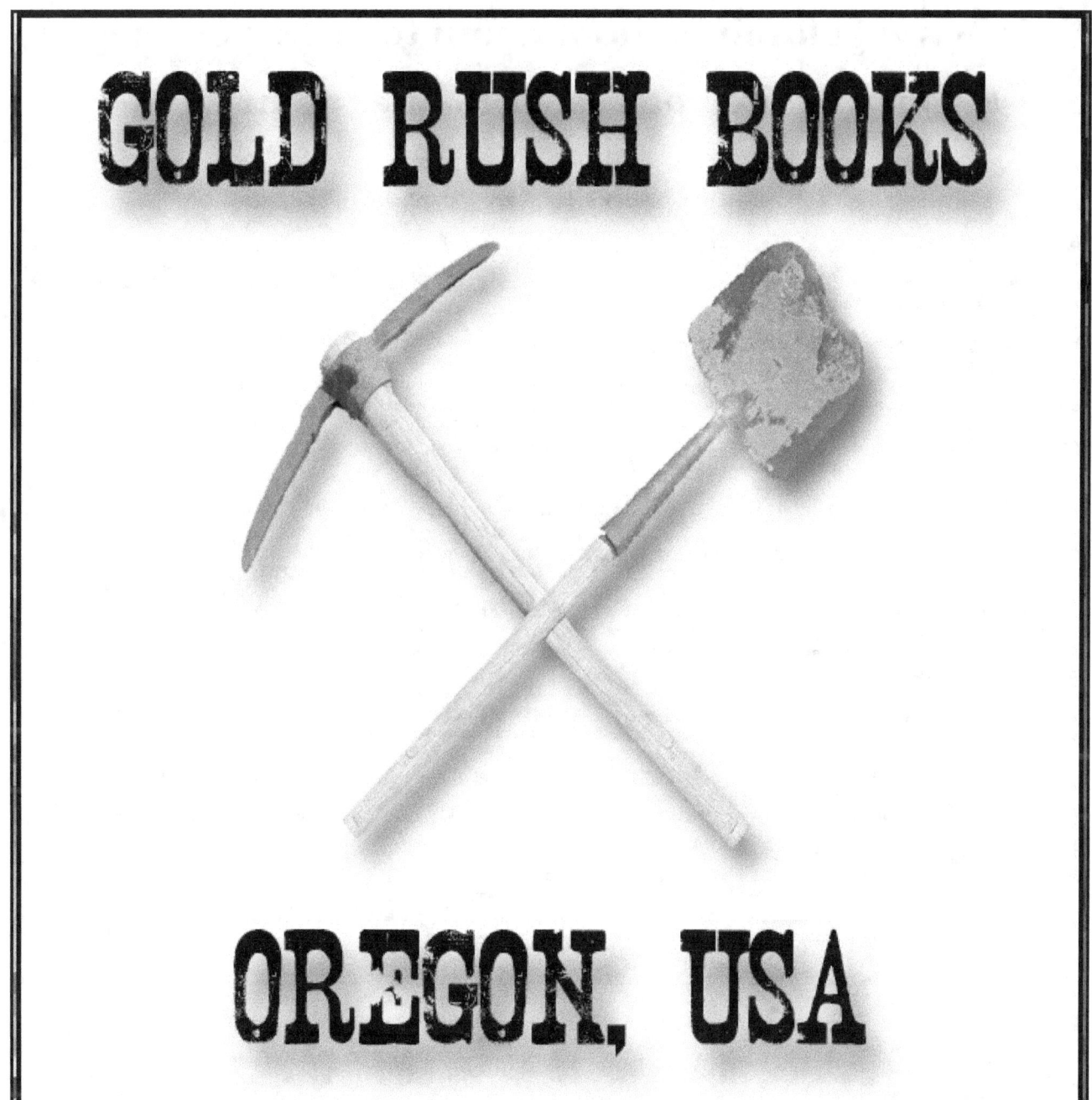

GOLD RUSH BOOKS

OREGON, USA

www.GoldMiningBooks.com

Books On Mining

Visit: www.goldminingbooks.com to order your copies or ask your favorite book seller to offer them.

Mining Books by Kerby Jackson

Gold Dust: Stories From Oregon's Mining Years - Oregon mining historian and prospector, Kerby Jackson, brings you a treasure trove of seventeen stories on Southern Oregon's rich history of gold prospecting, the prospectors and their discoveries, and the breathtaking areas they settled in and made homes. 5" X 8", 98 ppgs. Retail Price: $11.99

The Golden Trail: More Stories From Oregon's Mining Years - In his follow-up to "Gold Dust: Stories of Oregon's Mining Years", this time around, Jackson brings us twelve tales from Oregon's Gold Rush, including the story about the first gold strike on Canyon Creek in Grant County, about the old timers who found gold by the pail full at the Victor Mine near Galice, how Iradel Bray discovered a rich ledge of gold on the Coquille River during the height of the Rogue River War, a tale of two elderly miners on the hunt for a lost mine in the Cascade Mountains, details about the discovery of the famous Armstrong Nugget and others. 5" X 8", 70 ppgs. Retail Price: $10.99

Oregon Mining Books

Geology and Mineral Resources of Josephine County, Oregon - Unavailable since the 1970's, this important publication was originally compiled by the Oregon Department of Geology and Mineral Industries and includes important details on the economic geology and mineral resources of this important mining area in South Western Oregon. Included are notes on the history, geology and development of important mines, as well as insights into the mining of gold, copper, nickel, limestone, chromium and other minerals found in large quantities in Josephine County, Oregon. 8.5" X 11", 54 ppgs. Retail Price: $9.99

Mines and Prospects of the Mount Reuben Mining District - Unavailable since 1947, this important publication was originally compiled by geologist Elton Youngberg of the Oregon Department of Geology and Mineral Industries and includes detailed descriptions, histories and the geology of the Mount Reuben Mining District in Josephine County, Oregon. Included are notes on the history, geology, development and assay statistics, as well as underground maps of all the major mines and prospects in the vicinity of this much neglected mining district. 8.5" X 11", 48 ppgs. Retail Price: $9.99

The Granite Mining District - Notes on the history, geology and development of important mines in the well known Granite Mining District which is located in Grant County, Oregon. Some of the mines discussed include the Ajax, Blue Ribbon, Buffalo, Continental, Cougar-Independence, Magnolia, New York, Standard and the Tillicum. Also included are many rare maps pertaining to the mines in the area. 8.5" X 11", 48 ppgs. Retail Price: $9.99

Ore Deposits of the Takilma and Waldo Mining Districts of Josephine County, Oregon - The Waldo and Takilma mining districts are most notable for the fact that the earliest large scale mining of placer gold and copper in Oregon took place in these two areas. Included are details about some of the earliest large gold mines in the state such as the Llano de Oro, High Gravel, Cameron, Platerica, Deep Gravel and others, as well as copper mines such as the famous Queen of Bronze mine, the Waldo, Lily and Cowboy mines. This volume also includes six maps and 20 original illustrations. 8.5" X 11", 74 ppgs. Retail Price: $9.99

Metal Mines of Douglas, Coos and Curry Counties, Oregon - Oregon mining historian Kerby Jackson introduces us to a classic work on Oregon's mining history in this important re-issue of Bulletin 14C Volume 1, otherwise known as the Douglas, Coos & Curry Counties, Oregon Metal Mines Handbook. Unavailable since 1940, this important publication was originally compiled by the Oregon Department of Geology and Mineral Industries includes detailed descriptions, histories and the geology of over 250 metallic mineral mines and prospects in this rugged area of South West Oregon. 8.5" X 11", 158 ppgs. Retail Price: $19.99

Metal Mines of Jackson County, Oregon - Unavailable since 1943, this important publication was originally compiled by the Oregon Department of Geology and Mineral Industries includes detailed descriptions, histories and the geology of over 450 metallic mineral mines and prospects in Jackson County, Oregon. Included are such famous gold mining areas as Gold Hill, Jacksonville, Sterling and the Upper Applegate. **8.5" X 11", 220 ppgs. Retail Price: $24.99**

Metal Mines of Josephine County, Oregon - Oregon mining historian Kerby Jackson introduces us to a classic work on Oregon's mining history in this important re-issue of Bulletin 14C, otherwise known as the Josephine County, Oregon Metal Mines Handbook. Unavailable since 1952, this important publication was originally compiled by the Oregon Department of Geology and Mineral Industries includes detailed descriptions, histories and the geology of over 500 metallic mineral mines and prospects in Josephine County, Oregon. **8.5" X 11", 250 ppgs. Retail Price: $24.99**

Metal Mines of North East Oregon - Oregon mining historian Kerby Jackson introduces us to a classic work on Oregon's mining history in this important re-issue of Bulletin 14A and 14B, otherwise known as the North East Oregon Metal Mines Handbook. Unavailable since 1941, this important publication was originally compiled by the Oregon Department of Geology and Mineral Industries and includes detailed descriptions, histories and the geology of over 750 metallic mineral mines and prospects in North Eastern Oregon. **8.5" X 11", 310 ppgs. Retail Price: $29.99**

Metal Mines of North West Oregon - Oregon mining historian Kerby Jackson introduces us to a classic work on Oregon's mining history in this important re-issue of Bulletin 14D, otherwise known as the North West Oregon Metal Mines Handbook. Unavailable since 1951, this important publication was originally compiled by the Oregon Department of Geology and Mineral Industries and includes detailed descriptions, histories and the geology of over 250 metallic mineral mines and prospects in North Western Oregon. **8.5" X 11", 182 ppgs. Retail Price: $19.99**

Mines and Prospects of Oregon - Mining historian Kerby Jackson introduces us to a classic mining work by the Oregon Bureau of Mines in this important re-issue of The Handbook of Mines and Prospects of Oregon. Unavailable since 1916, this publication includes important insights into hundreds of gold, silver, copper, coal, limestone and other mines that operated in the State of Oregon around the turn of the 19th Century. Included are not only geological details on early mines throughout Oregon, but also insights into their history, production, locations and in some cases, also included are rare maps of their underground workings. **8.5" X 11", 314 ppgs. Retail Price: $24.99**

Lode Gold of the Klamath Mountains of Northern California and South West Oregon
(See California Mining Books)

Mineral Resources of South West Oregon - Unavailable since 1914, this publication includes important insights into dozens of mines that once operated in South West Oregon, including the famous gold fields of Josephine and Jackson Counties, as well as the Coal Mines of Coos County. Included are not only geological details on early mines throughout South West Oregon, but also insights into their history, production and locations. **8.5" X 11", 154 ppgs. Retail Price: $11.99**

Chromite Mining in The Klamath Mountains of California and Oregon
(See California Mining Books)

Southern Oregon Mineral Wealth - Unavailable since 1904, this rare publication provides a unique snapshot into the mines that were operating in the area at the time. Included are not only geological details on early mines throughout South West Oregon, but also insights into their history, production and locations. Some of the mining areas include Grave Creek, Greenback, Wolf Creek, Jump Off Joe Creek, Granite Hill, Galice, Mount Reuben, Gold Hill, Galls Creek, Kane Creek, Sardine Creek, Birdseye Creek, Evans Creek, Foots Creek, Jacksonville, Ashland, the Applegate River, Waldo, Kerby and the Illinois River, Althouse and Sucker Creek, as well as insights into local copper mining and other topics. **8.5" X 11", 64 ppgs. Retail Price: $8.99**

Geology and Ore Deposits of the Takilma and Waldo Mining Districts - Unavailable since the 1933, this publication was originally compiled by the United States Geological Survey and includes details on gold and copper mining in the Takilma and Waldo Districts of Josephine County, Oregon. The Waldo and Takilma mining districts are most notable for the fact that the earliest large scale mining of placer gold and copper in Oregon took place in these two areas. Included in this report are details about some of the earliest large gold mines in the state such as the Llano de Oro, High Gravel, Cameron, Platerica, Deep Gravel and others, as well as copper mines such as the famous Queen of Bronze mine, the Waldo, Lily and Cowboy mines. In addition to geological examinations, insights are also provided into the production, day to day operations and early histories of these mines, as well as calculations of known mineral reserves in the area. This volume also includes six maps and 20 original illustrations. **8.5" X 11", 74 ppgs. Retail Price: $9.99**

Gold Mines of Oregon - Oregon mining historian Kerby Jackson introduces us to a classic work on Oregon's mining history in this important re-issue of Bulletin 61, otherwise known as "Gold and Silver In Oregon". Unavailable since 1968, this important publication was originally compiled by geologists Howard C. Brooks and Len Ramp of the Oregon Department of Geology and Mineral Industries and includes detailed descriptions, histories and the geology of over 450 gold mines Oregon. Included are notes on the history, geology and gold production statistics of all the major mining areas in Oregon including the Klamath Mountains, the Blue Mountains and the North Cascades. While gold is where you find it, as every miner knows, the path to success is to prospect for gold where it was previously found. **8.5" X 11", 344 ppgs. Retail Price: $24.99**

Mines and Mineral Resources of Curry County Oregon - Originally published in 1916, this important publication on Oregon Mining has not been available for nearly a century. Included are rare insights into the history, production and locations of dozens of gold mines in Curry County, Oregon, as well as detailed information on important Oregon mining districts in that area such as those at Agness, Bald Face Creek, Mule Creek, Boulder Creek, China Diggings, Collier Creek, Elk River, Gold Beach, Rock Creek, Sixes River and elsewhere. Particular attention is especially paid to the famous beach gold deposits of this portion of the Oregon Coast. **8.5" X 11", 140 ppgs. Retail Price: $11.99**

Chromite Mining in South West Oregon - Originally published in 1961, this important publication on Oregon Mining has not been available for nearly a century. Included are rare insights into the history, production and locations of nearly 300 chromite mines in South Western Oregon. **8.5" X 11", 184 ppgs. Retail Price: $14.99**

Mineral Resources of Douglas County Oregon - Originally published in 1972, this important publication on Oregon Mining has not been available for nearly forty years. Included are rare insights into the geology, history, production and locations of numerous gold mines and other mining properties in Douglas County, Oregon. **8.5" X 11", 124 ppgs. Retail Price: $11.99**

Mineral Resources of Coos County Oregon - Originally published in 1972, this important publication on Oregon Mining has not been available for nearly forty years. Included are rare insights into the geology, history, production and locations of numerous gold mines and other mining properties in Coos County, Oregon. **8.5" X 11", 100 ppgs. Retail Price: $11.99**

Mineral Resources of Lane County Oregon - Originally published in 1938, this important publication on Oregon Mining has not been available for nearly seventy five years. Included are extremely rare insights into the geology and mines of Lane County, Oregon, in particular in the Bohemia, Blue River, Oakridge, Black Butte and Winberry Mining Districts. **8.5" X 11", 82 ppgs. Retail Price: $9.99**

Mineral Resources of the Upper Chetco River of Oregon: Including the Kalmiopsis Wilderness - Originally published in 1975, this important publication on Oregon Mining has not been available for nearly forty years. Withdrawn under the 1872 Mining Act since 1984, real insight into the minerals resources and mines of the Upper Chetco River has long been unavailable due to the remoteness of the area. Despite this, the decades of battle between property owners and environmental extremists over the last private mining inholding in the area has continued to pique the interest of those interested in mining and other forms of natural resource use. Gold mining began in the area in the 1850's and has a rich history in this geographic area, even if the facts surrounding it are little known. Included are twenty two rare photographs, as well as insights into the Becca and Morning Mine, the Emmly Mine (also known as Emily Camp), the Frazier Mine, the Golden Dream or Higgins Mine, Hustis Mine, Peck Mine and others. **8.5" X 11", 64 ppgs. Retail Price: $8.99**

Gold Dredging in Oregon - Originally published in 1939, this important publication on Oregon Mining has not been available for nearly seventy five years. Included are extremely rare insights into the history and day to day operations of the dragline and bucketline gold dredges that once worked the placer gold fields of South West and North East Oregon in decades gone by. Also included are details into the areas that were worked by gold dredges in Josephine, Jackson, Baker and Grant counties, as well as the economic factors that impacted this mining method. This volume also offers a unique look into the values of river bottom land in relation to both farming and mining, in how farm lands were mined, re-soiled and reclaimed after the dredges worked them. Featured are hard to find maps of the gold dredge fields, as well as rare photographs from a bygone era. **8.5" X 11", 86 ppgs. Retail Price: $8.99**

Quick Silver Mining in Oregon - Originally published in 1963, this important publication on Oregon Mining has not been available for over fifty years. This publication includes details into the history and production of Elemental Mercury or Quicksilver in the State of Oregon. **8.5" X 11", 238 ppgs. Retail Price: $15.99**

Mines of the Greenhorn Mining District of Grant County Oregon - Originally published in 1948, this important publication on Oregon Mining has not been available for over sixty five years. In this publication are rare insights into the mines of the famous Greenhorn Mining District of Grant County, Oregon, especially the famous Morning Mine. Also included are details on the Tempest, Tiger, Bi-Metallic, Windsor, Psyche, Big Johnny, Snow Creek, Banzette and Paramount Mines, as well as prospects in the vicinities in the famous mining areas of Mormon Basin, Vinegar Basin and Desolation Creek. Included are hard to find mine maps and dozens of rare photographs from the bygone era of Grant County's rich mining history. **8.5" X 11", 72 ppgs. Retail Price: $9.99**

Geology of the Wallowa Mountains of Oregon: Part I (Volume 1) - Originally published in 1938, this important publication on Oregon Mining has not been available for nearly seventy five years. Included are details on the geology of this unique portion of North Eastern Oregon. This is the first part of a two book series on the area. Accompanying the text are rare photographs and historic maps.8.5" X 11", **92 ppgs. Retail Price: $9.99**

Geology of the Wallowa Mountains of Oregon: Part II (Volume 2) - Originally published in 1938, this important publication on Oregon Mining has not been available for nearly seventy five years. Included are details on the geology of this unique portion of North Eastern Oregon. This is the first part of a two book series on the area. Accompanying the text are rare photographs and historic maps.8.5" X 11", **94 ppgs. Retail Price: $9.99**

Field Identification of Minerals For Oregon Prospectors - Originally published in 1940, this important publication on Oregon Mining has not been available for nearly seventy five years. Included in this volume is an easy system for testing and identifying a wide range of minerals that might be found by prospectors, geologists and rockhounds in the State of Oregon, as well as in other locales. Topics include how to put together your own field testing kit and how to conduct rudimentary tests in the field. This volume is written in a clear and concise way to make it useful even for beginners. 8.5" X 11", 158 ppgs. **Retail Price: $14.99**

The Bohemia Mining District of Oregon - Originally published in 1900, this important publication on Oregon Mining has not been available for over a century. Included in this volume are important insights into the famous Bohemia Mining District of Oregon, including the histories and locations of important gold mines in the area such as the Ophir Mine, Clarence, Acturas, Peek-a-boo, White Swan, Combination Mine, the Musick Mine, The California, White Ghost, The Mystery, Wall Street, Vesuvius, Story, Lizzie Bullock, Delta, Elsie Dora, Golden Slipper, Broadway, Champion Mine, Knott, Noonday, Helena, White Wings, Riverside and others. Also included are notes on the nearby Blue River Mining District. 8.5" X 11", 58 ppgs. **Retail Price: $9.99**

The Gold Fields of Eastern Oregon - Unavailable since 1900, this publication was originally compiled by the Baker City Chamber of Commerce Offering important insights into the gold mining history of Eastern Oregon, "The Gold Fields of Eastern Oregon" sheds a rare light on many of the gold mines that were operating at the turn of the 19th Century in Baker County and Grant County in North Eastern Oregon. Some of the areas featured include the Cable Cove District, Baisely-Elhorn, Granite, Red Boy, Bonanza, Susanville, Sparta, Virtue, Vaughn, Sumpter, Burnt River, Rye Valley and other mining districts. Included is basic information on not only many gold mines that are well known to those interested in Eastern Oregon mining history, but also many mines and prospects which have been mostly lost to the passage of time. Accompanying are numerous rare photos 8.5" X 11", 78 ppgs. **Retail Price: $10.99**

Gold Mining in Eastern Oregon - Originally published in 1938, this important publication on Oregon Mining has not been available for over a century. Included in this volume are important insights into the famous mining districts of Eastern Oregon during the late 1930's. Particular attention is given to those gold mines with milling and concentrating facilities in the Greenhorn, Red Boy, Alamo, Bonanza, Granite, Cable Cove, Cracker Creek, Virtue, Keating, Medical Springs, Sanger, Sparta, Chicken Creek, Mormon Basin, Connor Creek, Cornucopia and the Bull Run Mining Districts. Some of the mines featured include the Ben Harrison, North Pole-Columbia, Highland Maxwell, Baisley-Elkhorn, White Swan, Balm Creek, Twin Baby, Gem of Sparta, New Deal, Gleason, Gifford-Johnson, Cornucopia, Record, Bull Run, Orion and others. Of particular interest are the mill flow sheets and descriptions of milling operations of these mines. 8.5" X 11", 68 ppgs. **Retail Price: $8.99**

The Gold Belt of the Blue Mountains of Oregon - Originally published in 1901, this important publication on Oregon Mining has not been available for over a century. Included in this volume are rare insights into the gold deposits of the Blue Mountains of North East Oregon, including the history of their early discovery and early production. Extensive details are offered on this important mining area's mineralogy and economic geology, as well as insights into nearby gold placers, silver deposits and copper deposits. Featured are the Elkhorn and Rock Creek mining districts, the Pocahontas district, Auburn and Minersville districts, Sumpter and Cracker Creek, Cable Cove, the Camp Carson district, Granite, Alamo, Greenhorn, Robinsonville, the Upper Burnt River Valley and Bonanza districts, Susanville, Quartzburg, Canyon Creek, Virtue, the Copper Butte district, the North Powder River, Sparta, Eagle Creek, Cornucopia, Pine Creek, Lower Powder River, the Upper Snake River Canyon, Rye Valley, Lower Burnt River Valley, Mormon Basin, the Malheur and Clarks Creek districts, Sutton Creek and others. Of particular interest are important details on numerous gold mines and prospects in these mining districts, including their locations, histories, geology and other important information, as well as information on silver, copper and fire opal deposits. 8.5" X 11", 250 ppgs. **Retail Price: $24.99**

Mining in the Cascades Range of Oregon - Originally published in 1938, this important publication on Oregon Mining has not been available for over seventy five years. Included in this volume are rare insights into the gold mines and other types of metal mines in the Cascades Mountain Range of Oregon. Some of the important mining areas covered include the famous Bohemia Mining District, the North Santiam Mining District, Quartzville Mining District, Blue River Mining District, Fall Creek Mining District, Oakridge District, Zinc District, Buzzard-Al Sarena District, Grand Cove, Climax District and Barron Mining District. Of particular interest are important details on over 100 mines and prospects in these mining districts, including their locations, histories, geology and other important information. **8.5" X 11", 170 ppgs. Retail Price: $14.99**

Beach Gold Placers of the Oregon Coast - Originally published in 1934, this important publication on Oregon Mining has not been available for over 80 years. Included in this volume are rare insights into the beach gold deposits of the State of Oregon, including their locations, occurance, composition and geology. Of particular interest is information on placer platinum in Oregon's rich beach deposits. Also included are the locations and other information on some famous Oregon beach mines, including the Pioneer, Eagle, Chickamin, Iowa and beach placer mines north of the mouth of the Rogue River. **8.5" X 11", 60 ppgs. Retail Price: $8.99**

Idaho Mining Books

Gold in Idaho - Unavailable since the 1940's, this publication was originally compiled by the Idaho Bureau of Mines and includes details on gold mining in Idaho. Included is not only raw data on gold production in Idaho, but also valuable insight into where gold may be found in Idaho, as well as practical information on the gold bearing rocks and other geological features that will assist those looking for placer and lode gold in the State of Idaho. This volume also includes thirteen gold maps that greatly enhance the practical usability of the information contained in this small book detailing where to find gold in Idaho. **8.5" X 11", 72 ppgs. Retail Price: $9.99**

Geology of the Couer D'Alene Mining District of Idaho - Unavailable since 1961, this publication was originally compiled by the Idaho Bureau of Mines and Geology and includes details on the mining of gold, silver and other minerals in the famous Coeur D'Alene Mining District in Northern Idaho. Included are details on the early history of the Coeur D'Alene Mining District, local tectonic settings, ore deposit features, information on the mineral belts of the Osburn Fault, as well as detailed information on the famous Bunker Hill Mine, the Dayrock Mine, Galena Mine, Lucky Friday Mine and the infamous Sunshine Mine. This volume also includes sixteen hard to find maps. **8.5" X 11", 70 ppgs. Retail Price: $9.99**

The Gold Camps and Silver Cities of Idaho - Originally published in 1963, this important publication on Idaho Mining has not been available for nearly fifty years. Included are rare insights into the history of Idaho's Gold Rush, as well as the mad craze for silver in the Idaho Panhandle. Documented in fine detail are the early mining excitements at Boise Basin, at South Boise, in the Owyhees, at Deadwood, Long Valley, Stanley Basin and Robinson Bar, at Atlanta, on the famous Boise River, Volcano, Little Smokey, Banner, Boise Ridge, Hailey, Leesburg, Lemhi, Pearl, at South Mountain, Shoup and Ulysses, Yellow Jacket and Loon Creek. The story follows with the appearance of Chinese miners at the new mining camps on the Snake River, Black Pine, Yankee Fork, Bay Horse, Clayton, Heath, Seven Devils, Gibbonsville, Vienna and Sawtooth City. Also included are special sections on the Idaho Lead and Silver mines of the late 1800's, as well as the mining discoveries of the early 1900's that paved the way for Idaho's modern mining and mineral industry. Lavishly illustrated with rare historic photos, this volume provides a one of a kind documentary into Idaho's mining history that is sure to be enjoyed by not only modern miners and prospectors who still scour the hills in search of nature's treasures, but also those enjoy history and tromping through overgrown ghost towns and long abandoned mining camps. **8.5" X 11", 186 ppgs. Retail Price: $14.99**

Ore Deposits and Mining in North Western Custer County Idaho - Unavailable since 1913, this important publication was originally published by the Us Department of the Interior and has been unavailable for a century. Included are fine details on the geology, geography, gold placers and gold and silver bearing quartz veins of the mining region of North West Custer County, Idaho. Of particular interest is a rare look at the mines and prospects of the region, including those such as the Ramshorn Mine, SkyLark, Riverview, Excelsior, Beardsley, Pacific, Hoosier, Silver Brick, Forest Rose and dozens of others in the Bay Horse Mining District. Also covered are the mines of the Yankee Fork District such as the Lucky Boy, Badger, Black, Enterprise, Charles Dickens, Morrison, Golden Sunbeam, Montana, Golden Gate and others, as well as those in the Loon Mining District. **8.5" X 11", 126 ppgs. Retail Price: $12.99**

Gold Rush To Idaho - Unavailable since 1963, this important publication was originally published by the Idaho Bureau of Mines and has been unavailable for 50 years. "Gold Rush To Idaho" revisits the earliest years of the discovery of gold in Idaho Territory and introduces us to the conditions that the pioneer gold seekers met when they blazed a trail through the wilderness of Idaho's mountains and discovered the precious yellow metal at Oro Fino and Pierce. Subsequent rushes followed at places like Elk City, Newsome, Clearwater Station, Florence, Warrens and elsewhere. Of particular interest is a rare look at the hardships that the first miners in Idaho met with during their day to day existences and their attempts to bring law and order to their mining camps. 8.5" X 11", 88 ppgs. Retail Price: $9.99

The Geology and Mines of Northern Idaho and North Western Montana - Unavailable since 1909, this important publication was originally published by the Us Department of the Interior and has been unavailable for a century. Included are fine details on the geology and geography of the mining regions of Northern Idaho and North Western Montana. Of particular interest is a rare look at the mines and prospects of the region, including those in the Pine Creek Mining District, Lake Pend Oreille district, Troy Mining District, Sylvanite District, Cabinet Mining District, Prospect Mining District and the Missoula Valley. Some of the mines featured include the Iron Mountain, Silver Butte, Snowshoe, Grouse Mountain Mine and others. 8.5" X 11", 142 ppgs. Retail Price: $12.99

Mining in the Alturas Quadrangle of Blaine County Idaho - Unavailable since 1922, this important publication was originally published by the Idaho Bureau of Mines and has been unavailable for ninety years. Topics include the geology, rock formations and the formation of ore deposits in this important mining area of Idaho. Of particular focus is information on the local geology, quartz veins and ore deposits of this portion of Idaho. Included are hard to find details, including the descriptions and locations of numerous gold and silver mines in the area including the Silver King, Pilgrim, Columbia, Lone Jack, Sunbeam, Pride of the West, Lucky Boy, Scotia, Atlanta, Beaver-Bidwell and others mines and prospects. 8.5" X 11", 56 ppgs. Retail Price: $8.99

Mining in Lemhi County Idaho - Originally published in 1913, this important book on Idaho Mining has not been available to miners for over a century. Included are rare insights into hundreds of gold, silver, copper and other mines in this famous Idaho mining area. Details include the locations, geology, history, production and other facts of the mines of this region, not only gold and silver hardrock mines, but also gold placer mines, lead-silver deposits, copper mines, cobalt-nickel deposits, tungsten and tin mines . It is lavishly illustrated with hard to find photos of the period and rare mining maps. Some of the vicinities featured include the Nicholia Mining District, Spring Mountain District, Texas District, Blue Wing District, Junction District, McDevitt District, Pratt Creek, Eldorado District, Kirtley Creek, Carmen Creek, Gibbonsville, Indian Creek, Mineral Hill District, Mackinaw, Eureka District, Blackbird District, YellowJacket District, Gravel Range District, Junction District, Parker Mountain and other mining districts. 8.5" X 11", 226 ppgs. Retail Price: $19.99

Utah Mining Books

Fluorite in Utah - Unavailable since 1954, this publication was originally compiled by the USGS, State of Utah and U.S. Atomic Energy Commission and details the mining of fluorspar, also known as fluorite in the State of Utah. Included are details on the geology and history of fluorspar (fluorite) mining in Utah, including details on where this unique gem mineral may be found in the State of Utah. 8.5" X 11", 60 ppgs. Retail Price: $8.99

California Mining Books

The Tertiary Gravels of the Sierra Nevada of California - Mining historian Kerby Jackson introduces us to a classic mining work by Waldemar Lindgren in this important re-issue of The Tertiary Gravels of the Sierra Nevada of California. Unavailable since 1911, this publication includes details on the gold bearing ancient river channels of the famous Sierra Nevada region of California. 8.5" X 11", 282 ppgs. Retail Price: $19.99

The Mother Lode Mining Region of California - Unavailable since 1900, this publication includes details on the gold mines of California's famous Mother Lode gold mining area. Included are details on the geology, history and important gold mines of the region, as well as insights into historic mining methods, mine timbering, mining machinery, mining bell signals and other details on how these mines operated. Also included are insights into the gold mines of the California Mother Lode that were in operation during the first sixty years of California's mining history. 8.5" X 11", 176 ppgs. Retail Price: $14.99

Lode Gold of the Klamath Mountains of Northern California and South West Oregon - Unavailable since 1971, this publication was originally compiled by Preston E. Hotz and includes details on the lode mining districts of Oregon and California's Klamath Mountains. Included are details on the geology, history and important lode mines of the French Gulch, Deadwood, Whiskeytown, Shasta, Redding, Muletown, South Fork, Old Diggings, Dog Creek (Delta), Bully Choop (Indian Creek), Harrison Gulch, Hayfork, Minersville, Trinity Center, Canyon Creek, East Fork, New River, Denny, Liberty (Black Bear), Cecilville, Callahan, Yreka, Fort Jones and Happy Camp mining districts in California, as well as the Ashland, Rogue River, Applegate, Illinois River, Takilma, Greenback, Galice, Silver Peak, Myrtle Creek and Mule Creek districts of South Western Oregon. Also included are insights into the mineralization and other characteristics of this important mining region. 8.5" X 11", 100 ppgs. Retail Price: $10.99

Mines and Mineral Resources of Shasta County, Siskiyou County, Trinity County: California - Unavailable since 1915, this publication was originally compiled by the California State Mining Bureau and includes details on the gold mines of this area of Northern California. Also included are insights into the mineralization and other characteristics of this important mining region, as well as the location of historic gold mines. **8.5″ X 11″, 204 ppgs. Retail Price: $19.99**

Geology of the Yreka Quadrangle, Siskiyou County, California - Unavailable since 1977, this publication was originally compiled by Preston E. Hotz and includes details on the geology of the Yreka Quadrangle of Siskiyou County, California. Also included are insights into the mineralization and other characteristics of this important mining region. **8.5″ X 11″, 78 ppgs. Retail Price: $7.99**

Mines of San Diego and Imperial Counties, California - Originally published in 1914, this important publication on California Mining has not been available for a century. This publication includes important information on the early gold mines of San Diego and Imperial County, which were some of the first gold fields mined in California by early Spanish and Mexican miners before the 49ers came on the scene. Included are not only details on early mining methods in the area, production statistics and geological information, but also the location of the early gold mines that helped make California "The Golden State". Also included are details on the mining of other minerals such as silver, lead, zinc, manganese, tungsten, vanadium, asbestos, barite, borax, cement, clay, dolomite, fluospar, gem stones, graphite, marble, salines, petroleum, stronium, talc and others. **8.5″ X 11″, 116 ppgs. Retail Price: $12.99**

Mines of Sierra County, California - Unavailable since 1920, this publication was originally compiled by the California State Mining Bureau and includes details on the gold mines of Sierra County, California. Also included are insights into the mineralization and other characteristics of this important mining region, as well as the location of historic gold mines. **8.5″ X 11″, 156 ppgs. Retail Price: $19.99**

Mines of Plumas County, California - Unavailable since 1918, this publication was originally compiled by the California State Mining Bureau and includes details on the gold mines of Plumas County, California. Also included are insights into the mineralization and other characteristics of this important mining region, as well as the location of historic gold mines. **8.5″ X 11″, 200 ppgs. Retail Price: $19.99**

Mines of El Dorado, Placer, Sacramento and Yuba Counties, California - Originally published in 1917, this important publication on California Mining has not been available for nearly a century. This publication includes important information on the early gold mines of El Dorado County, Placer County, Sacramento County and Yuba County, which were some of the first gold fields mined by the Forty-Niners during the California Gold Rush. Included are not only details on early mining methods in the area, production statistics and geological information, but also the location of the early gold mines that helped make California "The Golden State". Also included are insights into the early mining of chrome, copper and other minerals in this important mining area. **8.5″ X 11″, 204 ppgs. Retail Price: $19.99**

Mines of Los Angeles, Orange and Riverside Counties, California - Originally published in 1917, this important publication on California Mining has not been available for nearly a century. This publication includes important information on the early gold mines of Los Angeles County, Orange County and Riverside County, which were some of the first gold fields mined in California by early Spanish and Mexican miners before the 49ers came on the scene. Included are not only details on early mining methods in the area, production statistics and geological information, but also the location of the early gold mines that helped make California "The Golden State". **8.5″ X 11″, 146 ppgs. Retail Price: $12.99**

Mines of San Bernadino and Tulare Counties, California - Originally published in 1917, this important publication on California Mining has not been available for nearly a century. This publication includes important information on the early gold mines of San Bernadino and Tulare County, which were some of the first gold fields mined in California by early Spanish and Mexican miners before the 49ers came on the scene. Included are not only details on early mining methods in the area, production statistics and geological information, but also the location of the early gold mines that helped make California "The Golden State". Also included are details on the mining of other minerals such as copper, iron, lead, zinc, manganese, tungsten, vanadium, asbestos, barite, borax, cement, clay, dolomite, fluospar, gem stones, graphite, marble, salines, petroleum, stronium, talc and others. **8.5″ X 11″, 200 ppgs. Retail Price: $19.99**

Chromite Mining in The Klamath Mountains of California and Oregon - Unavailable since 1919, this publication was originally compiled by J.S. Diller of the United States Department of Geological Survey and includes details on the chromite mines of this area of Northern California and Southern Oregon. Also included are insights into the mineralization and other characteristics of this important mining region, as well as the location of historic mines. Also included are insights into chromite mining in Eastern Oregon and Montana. **8.5″ X 11″, 98 ppgs. Retail Price: $9.99**

Mines and Mining in Amador, Calaveras and Tuolumne Counties, California - Unavailable since 1915, this publication was originally compiled by William Tucker and includes details on the mines and mineral resources of this important California mining area. Included are details on the geology, history and important gold mines of the region, as well as insights into other local mineral resources such as asbestos, clay, copper, talc, limestone and others. Also included are insights into the mineralization and other characteristics of this important portion of California's Mother Lode mining region. 8.5" X 11", 198 ppgs. Retail Price: $14.99

The Cerro Gordo Mining District of Inyo County California - Unavailable since 1963, this publication was originally compiled by the United States Department of Interior. Included are insights into the mineralization and other characteristics of this important mining region of Southern California. Topics include the mining of gold and silver in this important mining district in Inyo County, California, including details on the history, production and locations of the Cerro Gordo Mine, the Morning Star Mine, Estelle Tunnel, Charles Lease Tunnel, Ignacio, Hart, Crosscut Tunnel, Sunset, Upper Newtown, Newtown, Ella, Perseverance, Newsboy, Belmont and other silver and gold mines in the Cerro Gordo Mining District. This volume also includes important insights into the fossil record, geologic formations, faults and other aspects of economic geology in this California mining district. 8.5" X 11", 104 ppgs. Retail Price: $10.99

Mining in Butte, Lassen, Modoc, Sutter and Tehama Counties of California - Unavailable since 1917, this publication was originally compiled by the United States Department of Interior. Included are insights into the mineralization and other characteristics of this important mining region of California. Topics include the mining of asbestos, chromite, gold, diamonds and manganese in Butte County, the mining of gold and copper in the Hayden Hill and Diamond Mountain mining districts of Lassen County, the mining of coal, salt, copper and gold in the High Grade and Winters mining districts of Modoc County, gold mining in Sutter County and the mining of gold, chromite, manganese and copper in Tehama County. This volume also includes the production records and locations of numerous mines in this important mining region. 8.5" X 11", 114 ppgs. Retail Price: $11.99

Mines of Trinity County California - Originally published in 1965, this important publication on California Mining has not been available for nearly fifty years. This publication includes important information on mines and mining in Trinity County, California, as well insights into the mineralization and geology of this important mining area in Northern California. Included are extensive details on hardrock and placer gold mines and prospects, including charts showing the locations of these historic mines.. 8.5" X 11", 144 ppgs. Retail Price: $12.99

Mines of Kern County California - Originally published in 1962, this important publication on California Mining has not been available for nearly fifty years. This publication includes important information on mines and mining in Kern County, California, as well insights into the mineralization and geology of this important mining area in California. Included are extensive details on hardrock and placer gold mines and prospects, including charts showing the locations of these historic mines. 8.5" X 11", 398 ppgs. Retail Price: $24.99

Mines of Calaveras County California - Originally published in 1962, this important publication on California Mining has not been available for nearly fifty years. This publication includes important information on mines and mining in Calaveras County, California, as well insights into the mineralization and geology of this important mining area in Northern California. Included are extensive details on hardrock and placer gold mines and prospects, including charts showing the locations of these historic mines. 8.5" X 11", 236 ppgs. Retail Price: $19.99

Lode Gold Mining in Grass Valley California - Unavailable since 1940, this publication was originally compiled by the United States Department of Interior. Included are insights into the gold mineralization and other characteristics of this important mining region of Nevada County, California. This volume also includes important insights into the geologic formations, faults and other aspects of economic geology in this California mining district. Of particular interest are the fine details on many hardrock gold mines in the area, including their locations, histories, development and mineralization. Some of the mines featured include the Gold Hill Mine, Massachusetts Hill, Boundary, Peabody, Golden Center, North Star, Omaha, Lone Jack, Homeward Bound, Hartery, Wisconsin, Allison Ranch, Phoenix, Kate Hayes, W.Y.O.D., Empire, Rich Hill, Daisy Hill, Orleans, Sultana, Centennial, Conlin, Ben Franklin, Crown Point and many others. 8.5" X 11", 148 ppgs. Retail Price: $12.99

Lode Mining in the Alleghany District of Sierra County California - Unavailable since 1913, this publication was originally compiled by the United States Department of Interior. Included are insights into the mineralization and other characteristics of this important mining region of Sierra County. Included are details on the history, production and locations of numerous hardrock gold mines in this famous California area, including the Tightner Mine, Minnie D., Osceola, Eldorado, Twenty One, Sherman, Kenton, Oriental, Rainbow, Plumbago, Irelan, Gold Canyon, North Fork, Federal, Kate Hardy and others. This volume also includes important insights into the fossil record, geologic formations, faults and other aspects of economic geology in this California mining district. 8.5" X 11", 48 ppgs. Retail Price: $7.99

Six Months In The Gold Mines During The California Gold Rush - Unavailable since 1850, this important work is a first hand account of one "49'ers" personal experience during the great California Gold Rush, shedding important light on one of the most exciting periods in the history of not only California, but also the world. Compiled from journals written between 1847 and 1849 by E. Gould Buffum, a native of New York, "Six Months In The Gold Mines During The California Gold Rush" offers a rare look into the day to day lives of the people who came to California to work in her gold mines when the state was still a great frontier. 8.5" X 11", 290 ppgs. Retail Price: $19.99

Quartz Mines of the Grass Valley Mining District of California - Unavailable since 1867, this important publication has not been available since those days. This rare publication offers a short dissertation on the early hardrock mines in this important mining district in the California Mother Lode region between the 1850's and 1860's. Also included are hard to find details on the mineralization and locations of these mines, as well as how they were operated in those day. 8.5" X 11", 44 ppgs. Retail Price: $8.99

Alaska Mining Books

Ore Deposits of the Willow Creek Mining District, Alaska - Unavailable since 1954, this hard to find publication includes valuable insights into the Willow Creek Mining District near Hatcher Pass in Alaska. The publication includes insights into the history, geology and locations of the well known mines in the area, including the Gold Cord, Independence, Fern, Mabel, Lonesome, Snowbird, Schroff-O'Neil, High Grade, Marion Twin, Thorpe, Webfoot, Kelly-Willow, Lane, Holland and others. 8.5" X 11", 96 ppgs. Retail Price: $9.99

The Juneau Gold Belt of Alaska - Unavailable since 1906, this hard to find publication includes valuable insights into the gold mines around Juneau, Alaska. The publication includes important details into the history, geology and locations of the well known gold mines and prospects in the area, including those around Windham Bay, Holkham Bay, Port Snettisham, on Grindstone and Rhine Creeks, Gold Creek, Douglas Island, Salmon Creek, Lemon Creek, Nugget Creek, from the Mendenhall River to Berners Bay, McGinnis Creek, Montana Creek, Peterson Creek, Windfall Creek, the Eagle River, Yankee Basin, Yankee Curve, Kowee Creek and elsewhere. Not only are gold placer mines included, but also hardrock gold mines. 8.5" X 11", 224 ppgs. Retail Price: $19.99

Arizona Mining Books

Mines and Mining in Northern Yuma County Arizona - Originally published in 1911, this important publication on Arizona Mining has not been available for over a hundred years. Included are rare insights into the gold, silver, copper and quicksilver mines of Yuma County, Arizona together with hard to find maps and photographs. Some of the mines and mining districts featured include the Planet Copper Mine, Mineral Hill, the Clara Consolidated Mine, Viati Mine, Copper Basin prospect, Bowman Mine, Quartz King, Billy Mack, Carnation, the Wardwell and Osbourne, Valensuella Copper, the Mariquita, Colonial Mine, the French American, the New York-Plomosa, Guadalupe, Lead Camp, Mudersbach Copper Camp, Yellow Bird, the Arizona Northern (Salome Strike), Bonanza (Harqua Hala), Golden Eagle, Hercules, Socorro and others. 8.5" X 11", 144 ppgs. Retail Price: $11.99

The Aravaipa and Stanley Mining Districts of Graham County Arizona - Originally published in 1925, this important publication on Arizona Mining has not been available for nearly ninety years. Included are rare insights into the gold and silver mines of these two important mining districts, together with hard to find maps. 8.5" X 11", 140 ppgs. Retail Price: $11.99

Gold in the Gold Basin and Lost Basin Mining Districts of Mohave County, Arizona - This volume contains rare insights into the geology and gold mineralization of the Gold Basin and Lost Basin Mining Districts of Mohave County, Arizona that will be of benefit to miners and prospectors. Also included is a significant body of information on the gold mines and prospects of this portion of Arizona. This volume is lavishly illustrated with rare photos and mining maps. 8.5" X 11", 188 ppgs. Retail Price: $19.99

Mines of the Jerome and Bradshaw Mountains of Arizona - This important publication on Arizona Mining has not been available for ninety years. This volume contains rare insights into the geology and ore deposits of the Jerome and Bradshaw Mountains of Arizona that will be of benefit to miners and prospectors who work those areas. Included is a significant body of information on the mines and prospects of the Verde, Black Hills, Cherry Creek, Prescott, Walker, Groom Creek, Hassayampa, Bigbug, Turkey Creek, Agua Fria, Black Canyon, Peck, Tiger, Pine Grove, Bradshaw, Tintop, Humbug and Castle Creek Mining Districts. This volume is lavishly illustrated with rare photos and mining maps. 8.5" X 11", 218 ppgs. Retail Price: $19.99

The Ajo Mining District of Pima County Arizona - This important publication on Arizona Mining has not been available for nearly seventy years. This volume contains rare insights into the geology and mineralization of the Ajo Mining District in Pima County, Arizona and in particular the famous New Cornelia Mine. 8.5" X 11", 126 ppgs. Retail Price: $11.99

Mining in the Santa Rita and Patagonia Mountains of Arizona - Originally published in 1915, this important publication on Arizona Mining has not been available for nearly a century. Included are rare insights into hundreds of gold, silver, copper and other mines in this famous Arizona mining area. Details include the locations, geology, history, production and other facts of the mines of this region. **8.5" X 11", 394 ppgs. Retail Price: $24.99**

Mining in the Bisbee Quadrangle of Arizona - Originally published in 1906, this important publication on Arizona Mining has not been available for nearly a century. Included are rare insights into hundreds of gold, silver, copper and other mines in this famous Arizona mining area. Details include the locations, geology, history, production and other facts of the mines of this important mining region. **8.5" X 11", 188 ppgs. Retail Price: $14.99**

Montana Mining Books

A History of Butte Montana: The World's Greatest Mining Camp - First published in 1900 by H.C. Freeman, this important publication sheds a bright light on one of the most important mining areas in the history of The West. Together with his insights, as well as rare photographs of the periods, Harry Freeman describes Butte and its vicinity from its early beginnings, right up to its flush years when copper flowed from its mines like a river. At the time of publication, Butte, Montana was known worldwide as "The Richest Mining Spot On Earth" and produced not only vast amounts of copper, but also silver, gold and other metals from its mines. Freeman illustrates, with great detail, the most important mines in the vicinity of Butte, providing rare details on their owners, their history and most importantly, how the mines operated and how their treasures were extracted. Of particular interest are the dozens of rare photographs that depict mines such as the famous Anaconda, the Silver Bow, the Smoke House, Moose, Paulin, Buffalo, Little Minah, the Mountain Consolidated, West Greyrock, Cora, the Green Mountain, Diamond, Bell, Parnell, the Neversweat, Nipper, Original and many others. **8.5" X 11", 142 ppgs. Retail Price: $12.99**

The Butte Mining District of Montana - This important publication on Montana Mining has not been available for over a century. Included are rare insights into the gold, copper and silver mines of Butte, Montana together with hard to find maps and photographs. Some of the topics include the early history of gold, silver and copper mining in the Butte area, insight into the geology of its mining areas, the local distribution of gold, silver and copper ores, as well their composition and how to identify them. Also included are detailed facts about the mines in the Butte Mining District, including the famous Anaconda Mine, Gagnon, Parrot, Blue Vein, Moscow, Poulin, Stella, Buffalo, Green Mountain, Wake Up Jim, the Diamond-Bell Group, Mountain Consolidated, East Greyrock, West Greyrock, Snowball, Corra, Speculator, Adirondack, Miners Union, the Jessie-Edith May Group, Otisco, Iduna, Colorado, Lizzie, Cambers, Anderson, Hesperus, Preferencia and dozens of others. **8.5" X 11", 298 ppgs. Retail Price: $24.99**

Mines of the Helena Mining Region of Montana - This important publication on Montana Mining has not been available for over a century. Included are rare insights into the gold, copper and silver mines of the vicinity of Helena, Montana, including the Marysville Mining District, Elliston Mining District, Rimini Mining District, Helena Mining District, Clancy Mining District, Wickes Mining District, Boulder and Basin Mining Districts and the Elkhorn Mining District. Some of the topics include the early history of gold, silver and copper mining in the Helena area, insight into the geology of its mining areas, the local distribution of gold, silver and copper ores, as well their composition and how to identify them. Also included are detailed facts, history, geology and locations of over one hundred gold, silver and copper mines in the area . **8.5" X 11", 162 ppgs, Retail Price: $14.99**

Mines and Geology of the Garnet Range of Montana - This important publication on Montana Mining has not been available for over a century. Included are rare insights into the gold, copper and silver mines of the vicinity of this important mining area of Montana. Some of the topics include the early history of gold, silver and copper mining in the Garnet Mountains, insight into the geology of its mining areas, the local distribution of gold, silver and copper ores, as well their composition and how to identify them. Also included are detailed facts, history, geology and locations of numerous gold, silver and copper mines in the area . **8.5" X 11", 100 ppgs, Retail Price: $11.99**

Mines and Geology of the Philipsburg Quadrangle of Montana - This important publication on Montana Mining has not been available for over a century. Included are rare insights into the gold, copper and silver mines of the vicinity of this important mining area of Montana. Some of the topics include the early history of gold, silver and copper mining in the Philipsburg Quadrangle, insight into the geology of its mining areas, the local distribution of gold, silver and copper ores, as well their composition and how to identify them. Also included are detailed facts, history, geology and locations of over one hundred gold, silver and copper mines in the area **8.5" X 11", 290 ppgs, Retail Price: $24.99**

Geology of the Marysville Mining District of Montana - Included are rare insights into the mining geology of the Marysville Mining District. Some of the topics include the early history of gold, silver and copper mining in the area, insight into the geology of its mining areas, the local distribution of gold, silver and copper ores, as well their composition and how to identify them. Also included are detailed facts, history, geology and locations of gold, silver and copper mines in the area **8.5" X 11", 198 ppgs, Retail Price: $19.99**

The Geology and Mines of Northern Idaho and North Western Montana

See listing under Idaho.

Nevada Mining Books

The Bull Frog Mining District of Nevada - Unavailable since 1910, this publication was originally compiled by the United States Department of Interior. This volume also includes important insights into the geologic formations, faults and other aspects of economic geology in this Nevada mining district. Of particular interest are the fine details on many mines in the area, including their locations, histories, development and mineralization. Some of the mines featured include the National Bank Mine, Providence, Gibraltor, Tramps, Denver, Original Bullfrog, Gold Bar, Mayflower, Homestake-King and other mines and prospects. **8.5" X 11", 152 ppgs, Retail Price: $14.99**

History of the Comstock Lode - Unavailable since 1876, this publication was originally released by John Wiley & Sons. This volume also includes important insights into the famous Comstock Lode of Nevada that represented the first major silver discovery in the United States. During its spectacular run, the Comstock produced over 192 million ounces of silver and 8.2 million ounces of gold. Not only did the Comstock result in one of the largest mining rushes in history and yield immense fortunes for its owners, but it made important contributions to the development of the State of Nevada, as well as neighboring California. Included here are important details on not only the early development and history of the Comstock, but also rare early insight into its mines, ore and its geology.**8.5" X 11", 244 ppgs, Retail Price: $19.99**

Colorado Mining Books

Ores of The Leadville Mining District - Unavailable since 1926, this publication was originally compiled by the United States Department of Interior. This volume also includes important insights into the ores and mineralization of the Leadville Mining District in Colorado. Topics include historic ore prospecting methods, local geology, insights into ore veins and stockworks, the local trend and distribution of ore channels, reverse faults, shattered rock above replacement ore bodies, mineral enrichment in oxidized and sulphide zones and more. **8.5" X 11", 66 ppgs, Retail Price: $8.99**

Mining in Colorado - Unavailable since 1926, this publication was originally compiled by the United States Department of Interior. This volume also includes important insights into the mining history of Colorado from its early beginnings in the 1850's right up to the mid 1920's. Not only is Colorado's gold mining heritage included, but also its silver, copper, lead and zinc mining industry. Each mining area is treated separately, detailing the development of Colorado's mines on a county by county basis. **8.5" X 11", 284 ppgs, Retail Price: $19.99**

Gold Mining in Gilpin County Colorado - Unavailable since 1876, this publication was originally compiled by the Register Steam Printing House of Central City, Colorado. A rare glimpse at the gold mining history and early mines of Gilpin County, Colorado from their first discovery in the 1850's up to the "flush years" of the mid 1870's. Of particular interest is the history of the discovery of gold in Gilpin County and details about the men who made those first strikes. Special focus is given to the early gold mines and first mining districts of the area, many of which are not detailed in other books on Colorado's gold mining history. **8.5" X 11", 156 ppgs, Retail Price: $12.99**

Mining in the Gold Brick Mining District of Colorado - Important insights into the history of the Gold Brick Mining District, as well as its local geography and economic geology. Also included are the histories and locations of historic mines in this important Colorado Mining District, including the Cortland, Carter, Raymond, Gold Links, Sacramento, Bassick, Sandy Hook, Chronicle, Grand Prize, Chloride, Granite Mountain, Lucille, Gray Mountain, Hilltop, Maggie Mitchell, Silver Islet, Revenue, Roosevelt, Carbonate King and others. In addition to hardrock mining, are also included are details on gold placer mining in this portion of Colorado. **8.5" X 11", 140 ppgs, Retail Price: $12.99**

Washington Mining Books

The Republic Mining District of Washington - Unavailable since 1910, this important publication was originally published by the Washington Geologic Survey and has been unavailable for a century. Topics include the geology, rock formations and the formation of ore deposits in this important mining area of Washington State. Also included are hard to find details on the geology, history and locations of dozens of mines in the area. Some of the mines featured include the New Republic Mine, Ben Hur, Morning Glory, the South Republic Mine, Quilp, Surprise, Black Tail, Lone Pine, San Poil, Mountain Lion, Tom Thumb, Elcaliph and many others. **8.5" X 11", 94 ppgs, Retail Price: $10.99**

The Myers Creek and Nighthawk Mining Districts of Washington - Unavailable since 1911, this important publication was originally published by the Washington Geologic Survey and has been unavailable for a century. Topics include the geology, rock formations and the formation of ore deposits in these important mining areas of Washington State. Also included are hard to find details on the geology, history and locations of dozens of mines in the area. Some of the mines featured include the Grant Mine, Monterey, Nip and Tuck, Myers Creek, Number Nine, Neutral, Rainbow, Aztec, Crystal Butte, Apex, Butcher Boy, Molson, Mad River, Olentangy, Delate, Kelsey, Golden Chariot, Okanogan, Ohio, Forty-Ninth Parallel, Nighthawk, Favorite, Little Chopaka, Summit, Number One, California, Peerless, Caaba, Prize Group, Ruby, Mountain Sheep, Golden Zone, Rich Bar, Similkameen, Kimberly, Triune, Hiawatha, Trinity, Hornsilver, Maquae, Bellevue, Bullfrog, Palmer Lake, Ivanhoe, Copper World and many others.
8.5" X 11", 136 ppgs, Retail Price: $12.99

The Blewett Mining District of Washington - Unavailable since 1911, this important publication was originally published by the Washington Geologic Survey and has been unavailable for a century. Topics include the geology, rock formations and the formation of ore deposits in this important mining area of Washington State. Also included are hard to find details on the geology, history and locations of dozens of mines in the area. Some of the mines featured include the Washington Meteor, Alta Vista, Pole Pick, Blinn, North Star, Golden Eagle, Tip Top, Wilder, Golden Guinea, Lucky Queen, Blue Bell, Prospect, Homestake, Lone Rock, Johnson, and others. 8.5" X 11", 134 ppgs, Retail Price: $12.99

Silver Mining In Washington - Unavailable since 1955, this important publication was originally published by the Washington Geologic Survey. Featured are the hard to find locations and details pertaining to Washington's silver mines. 8.5" X 11", 180 ppgs, Retail Price: $15.99

The Mines of Snohomish County Washington - Unavailable since 1942, this important publication was originally published by the Washington Geologic Survey and has been unavailable for seventy years. Featured are details on a large number of gold, silver, copper, lead and other metallic mineral mines. Included are the locations of each historic mine, along with information on the commodity produced. 8.5" X 11", 98 ppgs, Retail Price: $10.99

The Mines of Chelan County Washington - Unavailable since 1943, this important publication was originally published by the Washington Geologic Survey and has been unavailable for seventy years. Featured are details on a large number of gold, silver, copper, lead and other metallic mineral mines. Included are the locations of each historic mine, along with information on the commodity. 8.5" X 11", 88 ppgs, Retail Price: $9.99

Metal Mines of Washington - Unavailable since 1921, this important publication was originally published by the Washington Geologic Survey and has been unavailable for nearly ninety years. Widely considered a masterpiece on the Washington Mining Industry, "Metal Mines of Washington" sheds light on the important details of Washington's early mining years. Featured are details on hundreds of gold, silver, copper, lead and other metallic mineral mines. Included are hard to find details on the mineral resources of this state, as well as the locations of historic mines. Lavishly illustrated with maps and historic photos and complete with a glossary to explain any technical terms found in the text, this is one of the most important works on mining in the State of Washington. No prospector or miner should be without it if they are interested in mining in Washington. 8.5" X 11", 396 ppgs, Retail Price: $24.99

Gem Stones In Washington - Unavailable since 1949, this important publication was originally published by the Washington Geologic Survey and has been unavailable since first published. Included are details on where to find naturally occurring gem stones in the State of Washington, including quartz crystal, amethyst, smoky quartz, milky quartz, agates, bloodstone, carnelian, chert, flint, jasper, onyx, petrified wood, opal, fire opal, hyalite and others. 8.5" X 11", 54 ppgs, Retail Price: $8.99

The Covada Mining District of Washington - Unavailable since 1913, this important publication was originally published by the Washington Geologic Survey and has been unavailable for a century. Topics include the geology, rock formations and the formation of ore deposits in this important mining area of Washington State. Also included are hard to find details on the geology, history and locations of dozens of mines in the area. Some of the mines featured include the Admiral, Advance, Algonkian, Big Bug, Big Chief, Big Joker, Black Hawk, Black Tail, Black Thorn, Captain, Cherokee Strip, Colorado, Dan Patch, Dead Shot, Etta, Good Ore, Greasy Run, Great Scott, Idora, IXL, Jay Bird, Kentucky Bell, King Solomon, Laurel, Laura S, Little Jay, Meteor, Neglected, Northern Light, Old Nell, Plymouth Rock, Polaris, Quandary, Reserve, Shoo Fly, Silver Plume, Three Pines, Vernie, White Rose and dozens of others. 8.5" X 11", 114 ppgs, Retail Price: $10.99

The Index Mining District of Washington - Unavailable since 1912, this important publication was originally published by the Washington Geologic Survey and has been unavailable for a century. Topics include the geology, rock formations and the formation of ore deposits in this important mining area of Washington State. Also included are hard to find details on the geology, history and locations of dozens of mines in the area. Some of the mines featured include the Sunset, Non-Pareil, Ethel Consolidated, Kittaning, Merchant, Homestead, Co-operative, Lost Creek, Uncle Sam, Calumet, Florence-Rae, Bitter Creek, Index Peacock, Gunn Peak, Helena, North Star, Buckeye. Copper Bell, Red Cross and others. 8.5" X 11", 114 ppgs, Retail Price: $11.99

Mining & Mineral Resources of Stevens County Washington - Unavailable since 1920, this important publication was originally published by the Washington Geologic Survey and has been unavailable for a century. Topics include the geology, rock formations and the formation of ore deposits in these important mining areas of Washington State. Also included are hard to find details on the geology, history and locations of hundreds of mines in the area. **8.5" X 11", 372 ppgs, Retail Price: $24.99**

The Mines and Geology of the Loomis Quadrangle Okanogan County, Washington - Unavailable since 1972, this important publication was originally published by the Washington Geologic Survey and has been unavailable for a century. Topics include the geology, rock formations and the formation of ore deposits in this important mining area of Washington State. Also included are hard to find details on the geology, history and locations of dozens of gold, copper, silver and other mines in the area. **8.5" X 11", 150 ppgs, Retail Price: $12.99**

The Conconully Mining District of Okanogan County Washington - Unavailable since 1973, this important publication was originally published by the Washington Geologic Survey and has been unavailable for a century. Topics include the geology, rock formations and the formation of ore deposits in this important mining area of Washington State, which also includes Salmon Creek, Blue Lake and Galena. Also included are hard to find details on the geology, mining history and locations of dozens of mines in the area. Some of the mines include Arlington, Fourth of July, Sonny Boy, First Thought, Last Chance, War Eagle-Peacock, Wheeler, Mohawk, Lone Star, Woo Loo Moo Loo, Keystone, Hughes, Plant-Callahan, Johnny Boy, Leuena, Gubser, John Arthur, Tough Nut, Homestake, Key and many others **8.5" X 11", 68 ppgs, Retail Price: $8.99**

Wyoming Mining Books

Mining in the Laramie Basin of Wyoming - Unavailable since 1909, this publication was originally compiled by the United States Department of Interior. Also included are insights into the mineralization and other characteristics of this important mining region, especially in regards to coal, limestone, gypsum, bentonite clay, cement, sand, clay and copper. **8.5" X 11", 104 ppgs, Retail Price: $11.99**

New Mexico Mining Books

The Mogollon Mining District of New Mexico - Unavailable since 1927, this important publication was originally published by the US Department of Interior and has been unavailable for 80 years. Topics include the geology, rock formations and the formation of ore deposits in this important mining area in New Mexico. Of particular focus is information on the history and production of the ore deposits in this area, their form and structure, vein filling, their paragenesis, origins and ore shoots, as well as oxidation and supergene enrichment. Also included are hard to find details, including the descriptions and locations of numerous gold, silver and other types of mines, including the Eureka, Pacific, South Alpine, Great Western, Enterprise, Buffalo, Mountain View, Floride, Gold Dust, Last Chance, Deadwood, Confidence, Maud S., Deep Down, Little Fanney, Trilby, Johnson, Alberta, Comet, Golden Eagle, Cooney, Queen, the Iron Crown, Eberle, Clifton, Andrew Jackson mine, Mascot and others. **8.5" X 11", 144 ppgs, Retail Price: $12.99**

The Percha Mining District of Kingston New Mexico - Unavailable since 1883, this important publication was originally published by the Kingston Tribune and has been unavailable for over one hundred and thirty five years. Having been written during the earliest years of gold and silver mining in the Percha Mining District, unlike other books on the subject, this work offers the unique perspective of having actually been written while the early mining history of this area was still being made. In fact, the work was written so early in the development of this area that many of the notable mines in the Percha District were less than a few years old and were still being operated by their original discoverers with the same enthusiasm as when they were first located. Included are hard to find details on the very earliest gold and silver mines of this important mining district near Kingston in Sierra County, New Mexico. **8.5" X 11", 68 ppgs, Retail Price: $9.99**

East Coast Mining Books

The Gold Fields of the Southern Appalachians - Unavailable since 1895, this important publication was originally published by the US Department of Interior and has been unavailable for nearly 120 years. Topics include the geology, rock formations and the formation of ore deposits in this important mining area of the American South. Of particular focus is information on the history and statistics of the ore deposits in this area, their form and structure and veins. Also included are details on the placer gold deposits of the region. The gold fields of the Georgian Belt, Carolinian Belt and the South Mountain Mining District of North Carolina are all treated in descriptive detail. Included are hard to find details, including the descriptions and locations of numerous gold mines in Georgia, North Carolina and elsewhere in the American South. Also included are details on the gold belts of the British Maritime Provinces and the Green Mountains. **8.5" X 11", 104 ppgs, Retail Price: $9.99**

Gold Rush Tales Series

Millions in Siskiyou County Gold - In this first volume of the "Gold Rush Tales" series, leading mining historian and editor Kerby Jackson, introduces us to the story of how millions of dollars worth of gold was discovered in Siskiyou County during the California Gold Rush. Lavishly illustrated with photos from the 19th Century, this hard to find information was first published in 1897 and sheds important light onto the gold rush era in Siskiyou County, California and the experiences of the men who dug for the gold and actually found it. 8.5" X 11", 82 ppgs, **Retail Price: $9.99**

The California Rand in the Days of '49 - In this second volume of the "Gold Rush Tales" series, leading mining historian and editor Kerby Jackson, introduces us to four tales from the California Gold Rush. Lavishly illustrated with photos from the 19th Century, this hard to find information was first published in 1890's and includes the stories of "California's Rand", details about Chinese miners, how one early miner named Baker struck it rich and also the story of Alphonzo Bowers, who invented the first hydraulic gold dredge. 8.5" X 11", 54 ppgs, **Retail Price: $9.99**

More Mining Books

Prospecting and Developing A Small Mine - Topics covered include the classification of varying ores, how to take a proper ore sample, the proper reduction of ore samples, alluvial sampling, how to understand geology as it is applied to prospecting and mining, prospecting procedures, methods of ore treatment, the application of drilling and blasting in a small mine and other topics that the small scale miner will find of benefit. 8.5" X 11", 112 ppgs, **Retail Price: $11.99**

Timbering For Small Underground Mines - Topics covered include the selection of caps and posts, the treatment of mine timbers, how to install mine timbers, repairing damaged timbers, use of drift supports, headboards, squeeze sets, ore chute construction, mine cribbing, square set timbering methods, the use of steel and concrete sets and other topics that the small underground miner will find of benefit. This volume also includes twenty eight illustrations depicting the proper construction of mine timbering and support systems that greatly enhance the practical usability of the information contained in this small book. 8.5" X 11", 88 ppgs. **Retail Price: $10.99**

Timbering and Mining - A classic mining publication on Hard Rock Mining by W.H. Storms. Unavailable since 1909, this rare publication provides an in depth look at American methods of underground mine timbering and mining methods. Topics include the selection and preservation of mine timbers, drifting and drift sets, driving in running ground, structural steel in mine workings, timbering drifts in gravel mines, timbering methods for driving shafts, positioning drill holes in shafts, timbering stations at shafts, drainage, mining large ore bodies by means of open cuts or by the "Glory Hole" system, stoping out ore in flat or low lying veins, use of the "Caving System", stoping in swelling ground, how to stope out large ore bodies, Square Set timbering on the Comstock and its modifications by California miners, the construction of ore chutes, stoping ore bodies by use of the "Block System", how to work dangerous ground, information on the "Delprat System" of stoping without mine timbers, construction and use of headframes and much more. This volume provides a reference into not only practical methods of mining and timbering that may be employed in narrow vein mining by small miners today, but also rare insights into how mines were being worked at the turn of the 19th Century. 8.5" X 11", 288 ppgs. **Retail Price: $24.99**

A Study of Ore Deposits For The Practical Miner - Mining historian Kerby Jackson introduces us to a classic mining publication on ore deposits by J.P. Wallace. First published in 1908, it has been unavailable for over a century. Included are important insights into the properties of minerals and their identification, on the occurrence and origin of gold, on gold alloys, insights into gold bearing sulfides such as pyrites and arsenopyrites, on gold bearing vanadium, gold and silver tellurides, lead and mercury tellurides, on silver ores, platinum and iridium, mercury ores, copper ores, lead ores, zinc ores, iron ores, chromium ores, manganese ores, nickel ores, tin ores, tungsten ores and others. Also included are facts regarding rock forming minerals, their composition and occurrences, on igneous, sedimentary, metamorphic and intrusive rocks, as well as how they are geologically disturbed by dikes, flows and faults, as well as the effects of these geologic actions and why they are important to the miner. Written specifically with the common miner and prospector in mind, the book will help to unlock the earth's hidden wealth for you and is written in a simple and concise language that anyone can understand. 8.5" X 11", 366 ppgs. **Retail Price: $24.99**

Mine Drainage - Unavailable since 1896, this rare publication provides an in depth look at American methods of underground mine drainage and mining pump systems. This volume provides a reference into not only practical methods of mining drainage that may be employed in narrow vein mining by small miners today, but also rare insights into how mines were being worked at the turn of the 19th Century. 8.5" X 11", 218 ppgs. **Retail Price: $24.99**

Fire Assaying Gold, Silver and Lead Ores - Unavailable since 1907, this important publication was originally published by the Mining and Scientific Press and was designed to introduce miners and prospectors of gold, silver and lead to the art of fire assaying. Topics include the fire assaying of ores and products containing gold, silver and lead; the sampling and preparation of ore for an assay; care of the assay office, assay furnaces; crucibles and scorifiers; assay balances; metallic ores; scorification assays; cupelling; parting' crucible assays, the roasting of ores and more. This classic provides a time honored method of assaying put forward in a clear, concise and easy to understand language that will make it a benefit to even beginners. 8.5" X 11", 96 ppgs. Retail Price: $11.99

Methods of Mine Timbering - Originally published in 1896, this important publication on mining engineering has not been available for nearly a century. Included are rare insights into historical methods of timbering structural support that were used in underground metal mines during the California that still have a practical application for the small scale hardrock miner of today. 8.5" X 11", 94 ppgs. Retail Price: $10.99

The Enrichment of Copper Sulfide Ores - First published in 1913, it has been unavailable for over a century. Topics include the definition and types of ore enrichment, the oxidation of copper ores, the precipitation of metallic sulfides. Also included are the results of dozens of lab experiments pertaining to the enrichment of sulfide ores that will be of interest to the practical hard rock mine operator in his efforts to release the metallic bounty from his mine's ore. 8.5" X 11", 92 ppgs. Retail Price: $9.99

A Study of Magmatic Sulfide Ores - Unavailable since 1914, this rare publication provides an in depth look at magmatic sulfide ores. Some of the topics included are the definition and classification of magmatic ores, descriptions of some magmatic sulfide ore deposits known at the time of publication including copper and nickel bearing pyrrohitic ore bodies, chalcopyrite-bornite deposits, pyritic deposits, magnetite-ileminite deposits, chromite deposits and magmatic iron ore deposits. Also included are details on how to recognize these types of ore deposits while prospecting for valuable hardrock minerals. 8.5" X 11", 138 ppgs. Retail Price: $11.99

The Cyanide Process of Gold Recovery - Unavailable since 1894 and released under the name "The Cyanide Process: Its Practical Application and Economical Results", this rare publication provides an in depth look at the early use of cyanide leaching for gold recovery from hardrock mine ores. This volume provides a reference into the early development and use of cyanide leaching to recover gold. 8.5" X 11", 162 ppgs. Retail Price: $14.99

California Gold Milling Practices - Unavailable since 1895 and released under the name "California Gold Practices", this rare publication provides an in depth look at early methods of milling used to reduce gold ores in California during the late 19th century. This volume provides a reference into the early development and use of milling equipment during the earliest years of the California Gold Rush up to the age of the Industrial Revolution. Much of the information still applies today and will be of use to small scale miners engaging in hardrock mining. 8.5" X 11", 104 ppgs. Retail Price: $10.99

Leaching Gold and Silver Ores With The Plattner and Kiss Processes - Mining historian Kerby Jackson introduces us to a classic mining publication on the evaluation and examination of mines and prospects by C.H. Aaron. First published in 1881, it has been unavailable for over a century and sheds important light on the leaching of gold and silver ores with the Plattner and Kiss processes. 8.5" X 11", 204 ppgs. Retail Price: $15.99

The Metallurgy of Lead and the Desilverization of Base Bullion - First published in 1896, it has been unavailable for over a century and sheds important light on the the recovery of silver from lead based ores. Some of the topics include the properties of lead and some of its compounds, lead ores such as galenite, anglesite, cerussite and others, the distribution of lead ores throughout the United States and the sampling and assaying of lead ores. Also covered is the metallurgical treatment of lead ores, as well as the desilverization of lead by the Pattinson Process and the Parkes Process. Hofman's text has long been considered one of the most important early works on the recovery of silver from lead based ores. 8.5" X 11", 452 ppgs. Retail Price: $29.99

Ore Sampling For Small Scale Miners - First published in 1916, it has been unavailable for over a century and sheds important light on historic methods of ore sampling in hardrock mines. Topics include how to take correct ore samples and the conditions that affect sampling, such as their subdivision and uniformity. Particular detail is given to methods of hand sampling ore bodies by grab sample, pipe sample and coning, as well as sampling by mechanical methods. Also given are insights into the screening, drying and grinding processes to achieve the most consistent sample results and much more. 8.5" X 11", 124 ppgs. Retail Price: $12.99

The Extraction of Silver, Copper and Tin from Ores - First published in 1896, it has been unavailable for over a century and sheds important light on how historic miners recovered silver, copper and tin from their mining operations. The book is split into three sections, including a discussion on the Lixiviation of Silver Ores, the mining and treatment of copper ores as practiced at Tharsis, Spain and the smelting of tin as it was practiced by metallurgists at Pulo Brani, Singapore. Also included is an overview and analysis of these historic metal recovery methods that will be of benefit to those interested in the extraction of silver, copper and tin from small mines. **8.5″ X 11″, 118 ppgs. Retail Price: $14.99**

The Roasting of Gold and Silver Ores - First published in 1880, it has been unavailable for over a century and sheds important light on how historic miners recovered gold and silver rom their mining operations. Topics include details on the most important silver and free milling gold ores, methods of desulphurization of ores, methods of deoxidation, the chlorination of ores, methods and details on roasting gold and silver ores, notes on furnaces and more. Also included are details on numerous methods of gold and silver recovery, including the Ottokar Hofman's Process, the Patera Process, Kiss Process, Augustin Process, Ziervogel Process and others. **8.5″ X 11″, 178 ppgs. Retail Price: $19.99**

The Examination of Mines and Prospects - First published in 1912, it has been unavailable for over a century and sheds important light on how to examine and evaluate hardrock mines, prospects and lode mining claims. Sections include Mining Examinations, Structural Geology, Structural Features of Ore Deposits, Primary Ores and their Distribution, Types of Primary Ore Deposits, Primary Ore Shoots, The Primary Alteration of Wall Rocks, Alterations by Surface Agencies, Residual Ores and their Distribution, Secondary Ores and Ore Shoots and Vein Outcrops. This hard to find information is a must for those who are interested in owning a mine or who already own a lode mining claim and wish to succeed at quartz mining. **8.5″ X 11″, 250 ppgs. Retail Price: $19.99**